Biochemical and Physiological Response During Oxidative Stress - From Invertebrates to Vertebrates

Edited by Marika Cordaro, Rosanna Di Paola and Roberta Fusco

Published in London, United Kingdom

Biochemical and Physiological Response During Oxidative Stress - From Invertebrates to Vertebrates
http://dx.doi.org/10.5772/10.5772/intechopen.1003333
Edited by Marika Cordaro, Rosanna Di Paola and Roberta Fusco

Contributors
Aamir Sohail, Aparajita Das, Caterina Faggio, Davide Di Paola, Elham Tavassoli Nejad, Fateme Souni, Gaffari Türk, George Albert Karikas, İbrahim Halil Güngör, Imran Ullah, Irum Nasir, Lifang Ma, Mahdi Banaee, Maria Trapali, Marika Cordaro, Marina A. Nakvasina, Marzieh Mahmoodi, Mehran Nouri, Mehroz Khan, Mohammad Amin Rajizadeh, Muhammad Ehsan Ul Haq, Muhammad Murtaza Hasnain, Naini Nishita Rao, Petros Karkalousos, Raza Sufyan, Reza Pourbabaki, Salvatore Cuzzocrea, Sarbani Giri, Seyfettin Gür, Shuang Shang, Valery G. Artyukhov, Yasemin Gündüztepe, Zainab Shateri

First published in London, United Kingdom, 2025 by IntechOpen
IntechOpen is the global imprint of INTECHOPEN LIMITED, registered in England and Wales, registration number: 11086078, 167-169 Great Portland Street, London, W1W 5PF, United Kingdom

For EU product safety concerns: IN TECH d.o.o., Prolaz Marije Krucifikse Kozulić 3, 51000 Rijeka, Croatia, info@intechopen.com or visit our website at intechopen.com.

British Library Cataloguing-in-Publication Data
A catalogue record for this book is available from the British Library

Biochemical and Physiological Response During Oxidative Stress - From Invertebrates to Vertebrates
Edited by Marika Cordaro, Rosanna Di Paola and Roberta Fusco
p. cm.

This title is part of the Biochemistry Book Series, Volume 61
Topic: Chemical Biology
Series Editor: Andrei Surguchov
Topic Editors: Murat Şentürk and Deniz Ekinci

Print ISBN 978-1-83769-775-5
Online ISBN 978-1-83769-774-8
eBook (PDF) ISBN 978-1-83769-776-2
ISSN 2632-0983

IntechOpen Book Series
Biochemistry
Volume 61

Aims and Scope of the Series

Biochemistry, the study of chemical transformations occurring within living organisms, impacts all of the life sciences, from molecular crystallography and genetics, to ecology, medicine and population biology. Biochemistry studies macromolecules - proteins, nucleic acids, carbohydrates and lipids –their building blocks, structures, functions and interactions. Much of biochemistry is devoted to enzymes, proteins that catalyze chemical reactions, enzyme structures, mechanisms of action and their roles within cells. Biochemistry also studies small signaling molecules, coenzymes, inhibitors, vitamins and hormones, which play roles in the life process. Biochemical experimentation, besides coopting the methods of classical chemistry, e.g., chromatography, adopted new techniques, e.g., X-ray diffraction, electron microscopy, NMR, radioisotopes, and developed sophisticated microbial genetic tools, e.g., auxotroph mutants and their revertants, fermentation, etc. More recently, biochemistry embraced the 'big data' omics systems. Initial biochemical studies have been exclusively analytic: dissecting, purifying and examining individual components of a biological system; in exemplary words of Efraim Racker, (1913 –1991) "Don't waste clean thinking on dirty enzymes." Today, however, biochemistry is becoming more agglomerative and comprehensive, setting out to integrate and describe fully a particular biological system. The 'big data' metabolomics can define the complement of small molecules, e.g., in a soil or biofilm sample; proteomics can distinguish all the proteins comprising e.g., serum; metagenomics can identify all the genes in a complex environment e.g., the bovine rumen.
This Biochemistry Series will address both the current research on biomolecules, and the emerging trends with great promise.

Meet the Series Editor

Andrei Surguchov, Ph.D., joined Baylor College of Medicine, Houston, TX, as a faculty member in 1992, where he studied the mechanisms of the genetic control of lipid metabolism. At the University of Utah, his research interests focused on cloning new genes encoding retinal proteins. He studied molecular and cellular mechanisms of neurodegenerative diseases and retinal degeneration at Washington University, St. Louis. Currently, his research focuses on the structure-function relationship of proteins involved in neurodegeneration and ocular diseases. Andrei Surguchov is an Editor-in-Chief at Biochemistry Research International and Associate Editor in several biomedical journals.

Meet the Volume Editors

Dr. Marika Cordaro graduated in Biology at the University of Messina in 2015. She completed his Ph.D. in Applied Biology and Experimental Medicine at the University of Messina in Italy in 2017. She is a young researcher in physiology, particularly interested in cell physiology and molecular mechanisms underlying inflammation. During her research activity, Dr. Cordaro produced about 190 articles that attest to a diffuse interest in nutrition and aging disorder-related. She is also interested in endocrine disruptors and their role in neurological deficits. Her research is involved in preclinical studies to discover physiological cellular responses that could be used as new potential therapeutic targets.

Prof. Rosanna Di Paola has a Ph.D. in Experimental Medicine, a Ph.D. in Biotechnology, Pharmacology and Clinical Pharmacology and specializes in Clinical Pathology. She is now a Full Professor of Biochemistry at the Department of Veterinary Sciences of the University of Messina in Italy. She is the author of over 417 publications in renowned journals with 16. 616 citations and an h-index of 65. Her areas of scientific interest are autoimmunity, acute and chronic inflammation and signaling pathway modulation. Thanks to her long experience in laboratory approaches, Professor Di Paola has developed and acquired various skills in analytical methodologies, including experimental animal models, histology, immunohistochemistry and molecular biology techniques for studying protein expression.

Dr. Fusco completed her Ph.D. in Applied Biology and Experimental Medicine at the University of Messina and Yale University School of Medicine. She worked as a researcher for studies on biochemical activities (in vitro primary cultures and in vivo experimental models) for the Epitech Group, a biological company. Now, she is an Assistant Professor of Biochemistry at the Chemical, Biological, Pharmaceutical and Environmental Science Department of the University of Messina in Italy. Dr. Fusco has a long-lasting expertise in the induction and management of in vitro and in vivo inflammation and oxidative stress models, which is documented by more than 100 papers published in international refereed journals. Her areas of scientific interest are the role of oxidative stress in autoimmunity, inflammation and neurodegenerative disorders; nutritional antioxidants; and modulation of signal pathways.

Contents

Preface

This book covers the involvement of oxidative stress in immune dysfunction, diabetes, cancer, and neurodegenerative diseases in a broad sense.

The text begins by discussing the impact of ROS on the immune system, showing how these molecules can interfere with lymphocyte functions by modulating cytotoxic activity, antibody production, and surface receptor expression. Then, it addresses ferroptosis, a new form of regulated cell death resulting from dysregulated lipid metabolism, and further discusses its implications in various pathological conditions.

The book further discusses the pathogenesis of type 2 diabetes, discussing how hyperglycemia-induced oxidative stress induces insulin resistance and complication developments characteristic of the disease. The role of ROS in carcinogenesis is also discussed, including the possible therapeutic applications of polyphenols acting as pro-oxidants inside tumor cells.

Another important topic covered in this book is the use of biomarkers as important tools for assessing exposure to xenobiotics and evaluating their effects on the environment and human health. The text concludes with a detailed analysis of toxic effects on human health due to metals, considering their ability to cause oxidative stress and further lead to various diseases.

The book covers several aspects of oxidative stress, from the basic mechanisms of its pathogenesis to its clinical applications, thus representing an important source for new prevention and treatment strategies.

Marika Cordaro
Department of Biomedical and Dental Sciences and Morphofunctional Imaging,
University of Messina,
Italy

Rosanna Di Paola
Department of Veterinary Sciences,
University of Messina,
Italy

Roberta Fusco
Department of Chemical, Biological, Pharmaceutical, and Environmental Sciences,
University of Messina,
Italy

Chapter 1

Dietary Phytochemical Index and Oxidative Balance Score Could Effect on Bone Mass Density

Marzieh Mahmoodi, Fateme Souni, Elham Tavassoli Nejad, Zainab Shateri and Mehran Nouri

Abstract

Osteopenia is defined as a condition characterized by reduced bone mineral density, influenced by various factors. It is understood that following a healthy lifestyle, including factors such as maintaining an adequate diet, exposure to sunlight, and regular physical activity, can play an important role in preventing some disorders while potentially causing others. Diet is one of the modifiable lifestyle factors that potentially plays an important role in either promoting or mitigating oxidative stress, depending on whether it contains oxidants or antioxidants. Assessing the intake of specific nutrients is crucial for bone health, and is significant through the evaluation of dietary indices. In summary, all reviewed findings suggest that adhering to a plant-based diet rich in phytochemicals and specific antioxidants may enhance the attainment of optimal bone mass and decrease the risk of osteopenia and osteoporosis.

Keywords: diet, phytochemical, oxidative balance score, bone mass density, osteopenia

1. Introduction

Osteopenia is recognized as a clinical disorder characterized by decreased bone mineralization [1]. This disease is associated with a decrease in bone mineral density (BMD), reflecting the underlying deterioration of bone microarchitecture [1], leading to fragility fractures [2]. According to the World Health Organization (WHO) guidelines, osteopenia is defined as a T-score of BMD between −1 and − 2.5, while osteoporosis is defined as a T-score of BMD less than −2.5 [1, 2]. The T-score is calculated by dividing the difference in BMD between the patient population and the young normal population by the standard deviation (SD) of the young normal population [2].

2. Prevalence

Currently, 34 million Americans have osteopenia, with its incidence rising as the population ages. Additionally, it has been observed that the risk of developing the disease increases with age [1]. In the United States, 54% of postmenopausal women have osteopenia. However, by age 80, this proportion decreases to 27%, with approximately 70% transitioning to osteoporosis. This trend suggests an increasing prevalence of osteoporosis with advancing age [1]. Asia has exhibited the lowest average T-scores among worldwide regions. In Australia, the incidence rate of osteopenia is 42% in men and 51% in women. In 2005, India also reported an overall incidence of 52% for osteopenia [1].

Osteopenia affects both women and men [1], but women are reported to be at four times higher risk than men [2]. In general, osteopenia affects 39.4% of women and 42.1% of postmenopausal women [3], because they are at a higher risk of developing osteopenia [2]. It has been reported that women have lower average and maximum bone mass compared to men, possibly due to hormonal changes, resulting in a 40–50% decrease in maximum bone mass in women [1]. Because BMD reduction is common in women, especially with the onset of menopause around the age of 50 [4]. There is a hypothesis suggesting that sex hormones protect bones against oxidative stress due to their antioxidant properties [4]. Oxidative stress plays a crucial role in the development of bone disorders, such as osteoporosis [5].

3. Risk factors

Risk factors for developing osteopenia and osteoporosis include family history, premature menopause, hypogonadism, amenorrhea or oligomenorrhea in the second and third decades, inadequate intake of calcium, chronic deficiency of vitamin D, smoking, excessive consumption of alcohol and soft drink, low physical activity, taking steroids, and other drugs such as heparin, valproic acid, proton pump inhibitors, and methotrexate [6].

3.1 Nutritional aspects of osteopenia

In addition to the factors mentioned, lifestyle factors also play a key role in influencing bone density [4]. Hence, it has been suggested that osteoporosis and osteopenia should be regarded as lifestyle-related disorders [7]. It can be understood that following a healthy lifestyle, including factors such as maintaining an adequate diet, exposure to sunlight, and regular physical activity, can play an important role in both preventing and contributing to these disorders [7]. Among lifestyle factors, the potential impact of nutrition on bone health has also been suggested [4]. In this regard, epidemiological findings underscore the influence of dietary habits on bone health. Recently, more attention has been devoted to investigating the relationship between nutrients and the risk of fracture and low bone density [8]. For example, nutrients essential for bone health, including calcium, vitamin D, magnesium, and vitamin K, play crucial roles in bone development and promotion of bone health [9]. Furthermore, adequate intake of milk and dairy products has been reported to have a potential protective effect on bone health [8, 10, 11], fruits, and vegetables [8, 12], and plant and animal dietary protein [8, 13, 14]. Evaluating the intake of these essential nutrients, which impact bone health, is important through the determination of dietary indices.

4. Dietary phytochemical index (DPI)

Diet is a modifiable lifestyle factor that may play an important role in either promoting or mitigating oxidative stress, depending on whether it contains oxidants or antioxidants [15]. For example, dietary compounds such as fruits, vegetables, whole grains, nuts, and legumes are known for their antioxidant properties, attributed to their high phytochemical content [15]. Phytochemicals are non-nutritive bioactive compounds of plant origin known for their wide range of potential health benefits [16, 17]. These compounds are classified into several categories based on their structure and chemical properties, and they are commonly found in fruits, vegetables, whole grains, nuts, and other plant-based foods [16]. The protective effects of phytochemicals against chronic disorders are attributed to their antioxidant and anti-inflammatory properties [18]. Therefore, considering the potential beneficial effects of phytochemicals, it is recommended to assess the intake of phytochemicals in the diet [19]. However, as it is not feasible to directly determine dietary phytochemical content, an alternative method called the dietary phytochemical index (DPI) has been proposed by McCarty [19, 20].

In this method, the amount of dietary phytochemical intake can be evaluated [20], which is defined as the percentage of calories from phytochemical-rich foods (DPI = daily energy from phytochemical-rich food) (kcal) /total daily energy intake (kcal) * 100) [16, 19]. DPI serves as an indicator of dietary phytochemical content and reflects overall diet quality [19]. Furthermore, in epidemiological studies, it can also serve as an indicator of adherence to a healthier diet and may indicate subsequent broader health benefits [16].

Therefore, a higher DPI is associated with better health status and a reduced risk of chronic disorders [15]. The findings have centered on evaluating the relationship between the score, dietary indices, and bone health [21]. Observational studies have investigated the association between adherence to a healthy diet, emphasizing the consumption of nutrient-rich food groups, as well as poor dietary habits, and changes in BMD [22–24], as adhering to high-quality and healthy dietary patterns has been shown to positively impact bone health [2]. For example, in Ghadiri et al.'s [2] case-control study involving postmenopausal women, a strong and positive relationship was reported between adherence to the healthy eating index (HEI) and dietary quality index (DQI) and bone health. In this regard, in the Framingham osteoporosis study [25] and Rotterdam study [26], it has been shown that adherence to nutrient-dense dietary patterns with emphasis on fruit and vegetable consumption has a positive effect on BMD levels [27]. In Brondani et al.'s [27] systematic review and meta-analysis study, it was emphasized that adherence to dietary patterns rich in fruits and vegetables reduces the risk of hip fracture. Consuming more fruits and vegetables leads to adherence to a healthier dietary pattern and reduces sugar and fat intake, which positively affects BMD levels [27]. Moreover, research has shown that consuming Mediterranean food groups, including fruits and vegetables, is essential for maintaining bone health [28]. So, it appears that adhering to dietary patterns or consuming food groups rich in phytochemicals may effectively contribute to maintaining bone health and preventing osteopenia and osteoporosis.

Various mechanisms can be considered to explore the association between phytochemical-rich foods and bone health. The plant foods included in the DPI are the primary sources of soluble and insoluble fibers, lignans, phenolic acids, stilbenes, and flavonoids, which prevent chronic disorders through the action of their bioactive metabolites [15]. According to the study of Mirmiran et al. [29] and Vincent et al. [30], individuals' higher dietary phytochemical score is associated with higher consumption of fiber, carotenoids, vitamin C, other antioxidants, and phytochemicals. Phytochemicals are the main components of the diet with potential antioxidant and anti-inflammatory properties [19]. Studies have demonstrated that the antioxidant and anti-inflammatory effects, as well as the impact on bone health, of phytochemicals such as tannins, polyphenols, and flavonoids [31]. Furthermore, phytochemical compounds have also been reported to exhibit proestrogenic activity [31, 32]. This effect is associated with osteoclast apoptosis, which arises from estrogen receptor (ER) activation and the regulation of the ratio between receptor activator of nuclear factor kappa-B ligand (RANKL) and osteoprotegerin. These factors are the main regulators of osteoclastogenesis [32].

Additionally, as mentioned, consuming more fruits and vegetables as rich sources of phytochemicals has the potential to increase BMD levels, reduce bone loss and fracture rates, and ultimately improve bone health [33]. According to the findings, phytochemicals found in fruits and vegetables, such as isoflavones, polyphenols, lycopene, genistein, and carotenoids, contribute to the strengthening of BMD and the promotion of osteogenesis. This phenomenon is associated with the beneficial effects of phytochemicals, such as the inhibition of osteoclast formation, reduction of cytokine damage, enhancement of bone cell growth, and decrease in bone resorption markers [34]. Similarly, greater adherence to the Dietary Approaches to Stop Hypertension (DASH) diet, which emphasizes fruits, vegetables, and other phytochemical-rich sources, is positively associated with bone health [33]. Since fruits and vegetables are the main components of the phytochemical index [16], different mechanisms have been explained for the positive effects of fruits and vegetables. It has been reported that fruits and vegetables are rich in alkaline ions, such as calcium, magnesium, and potassium. These ions create alkaline conditions that largely counteract calcisuria derived from the diet [12, 33, 35]. Moreover, magnesium positively affects bone health by enhancing osteoblast survival, alkaline phosphatase activity, and osteocalcin levels. Vitamin K, found in fruits and vegetables, also plays a potential role in bone health by strengthening the gamma-carboxylation of osteocalcin, alkaline phosphatase activity, insulin-like growth factor 1 (IGF-1), and regulating extracellular matrix mineralization [33]. In addition, the intake of polyphenols, carotenoids, tocopherols, vitamin K, and glutathione, which have antioxidant effects, leads to the up-regulation of runt-related transcription factor 2 (Runx2), Osterix, and insulin-like growth factor 1 (IGF-1), and increased expression of lysyl oxidase, all of which contribute positively to bone health [33]. Finally, vitamin C intake, with its antioxidant effect, has the potential to enhance the activity of collagen synthase types 1 and 3, thus offering protection against bone fracture and osteoporosis [33].

4.1 Oxidative balance score (OBS)

Another environmental factor affecting bone density is oxidative stress, characterized by an imbalance between the excessive production of reactive oxygen species (ROS) and the antioxidant defense system [36, 37]. Oxidative stress damages proteins, lipids, and DNA, thereby adversely affecting overall health [38]. Oxidative stress increases bone remodeling turnover and bone mass loss by enhancing the inflammatory response and regulating osteoblast survival and differentiation. Conversely, antioxidants play a significant role in reducing bone mass loss by inhibiting oxidative stress [39]. As shown by Domazetovic et al. [40], the consumption of fruits and vegetables with high levels of antioxidants has resulted in a reduction in bone mass loss and, consequently, a decreased risk of osteoporosis in postmenopausal women. Epidemiological findings have also reported a positive relationship between the intake of antioxidants, such as vitamin E, vitamin C, flavonoids, and carotenoids, and bone mineral density (BMD) in both premenopausal and postmenopausal women [39].

Findings have also shown that dietary intake of antioxidants mitigates the adverse effects caused by oxidative stress [38]. Evidence suggests a relationship between menopause-related estrogen deprivation, oxidative stress-induced bone loss, and the development of postmenopausal osteoporosis [36]. However, the results are contradictory, and no observational studies have demonstrated a direct and significant relationship between pro-oxidant exposures and postmenopausal osteoporosis [36]. Several randomized controlled trial (RCT) studies have also reported a significant effect of exposure to antioxidants on bone health markers in postmenopausal women [36]. In this regard, the confirmed potential role of oxidative stress in many chronic disorders, such as osteoporosis, is evident [36].

Hence, the OBS index was reported for the first time by Von Hoydonck et al. Still, this index only included two antioxidants (vitamin C and beta-carotene) and one pro-oxidant factor (iron). Later, this index was proposed by Goodman et al., which includes nine antioxidants and three pro-oxidant factors [37]. Goodman et al. [36, 41] have proposed OBS as a combined measure of anti- and pro-oxidant exposure levels. Thus, a higher OBS indicates a higher exposure to antioxidants than pro-oxidants [36, 42]. This index is a practical and useful tool for checking individuals' antioxidant and pro-oxidant balance, including dietary and environmental factors [43]. Therefore, it shows a complete and comprehensive assessment of individuals' oxidative stress status and antioxidant capacity compared to markers alone [43].

Hence, several epidemiological studies have used this index to predict the risk of chronic disorders, including osteoporosis [36], because OBS is inversely associated with chronic oxidative stress-related disorders [36]. As Zhou et al. [44] reported in a systematic review and meta-analysis study, postmenopausal osteoporosis is associated with increased levels of oxidative stress and decreased antioxidant defense systems. Moreover, according to the findings, the positive effect of dietary and non-dietary factors included in the calculation of OBS, such as smoking, high intake of saturated fatty acid (SFA), low physical activity, or dietary intake of fiber, vitamin C, vitamin E, folate, carotenoids, zinc, and selenium has been reported to increase the risk of osteoporosis [36, 40, 45]. Various mechanisms have been described to investigate this association. Oxidative stress, by inhibiting osteoblast activity, inducing osteoclast differentiation, and osteocyte apoptosis, leads to an imbalance between bone formation and bone resorption [36]. Oxidative stress-induced osteocyte apoptosis leads to an increase in the expression of pro-inflammatory biomarkers, including tumor necrosis factor-alpha (TNF-a), interleukin (IL)-1 beta (IL-1β), and IL-6, and finally, an increase in the level of ROS associated with an increase in the expression of fibroblast growth factor-23 (FGF23) and a decrease in mineralization [46].

In contrast, antioxidants can potentially prevent and control oxidative stress conditions and oxidative stress-induced negative effects on bone metabolism. It has been shown that antioxidants inhibit osteoclastogenesis and activate the differentiation and osteogenic activity of osteoblasts and the formation of bone matrix, ultimately maintaining the sufficient activity of osteoblasts and osteoclasts and inhibiting oxidative stress-induced bone loss [46]. The findings have also emphasized the relationship between consuming foods rich in antioxidants and bone protection [46, 47]. For example, the effect of vitamins E and C on inhibition of osteoclast activity and stimulation of osteoblast activation has been reported. The favorable effect of several natural antioxidants due to their role in maintaining normal levels of glutathione (GSH) in enhancing osteogenesis and the mineralization process via the antioxidant and non-redox-regulated mechanisms of GSH has also been shown [46]. In this view, higher OBS with higher exposure to antioxidants than pro-oxidants may potentially prevent or reverse oxidative stress-induced bone loss [36].

5. Conclusions

In summary, all reviewed findings suggest that adherence to a plant-based diet with a high content of phytochemicals and certain antioxidants may enhance the achievement of optimal bone mass and reduce the risk of osteopenia and osteoporosis. Therefore, the consumption of food groups or adherence to dietary patterns rich in antioxidants and phytochemicals is suggested as an alternative tool to strengthen bone health. However, it is recommended that more studies be carried out for a comprehensive and detailed description of the involved molecular mechanisms for this purpose (**Table 1**).

Study (Year)	Country	Objective	Type of study	Key findings
Tucker et al. [25]	USA	Examine the association between dietary patterns and bone mineral density	Cohort	1. A protective effect of higher consumption of fruit and vegetables on BMD in men. 2. A detrimental effect of higher consumption of candy on BMD in men and women.
Macdonald et al. [24]	Scotland	Examine the association between calcium, alcohol, fruit, vegetable, and fatty acid with bone loss	Longitudinal	Protective effect of fruit and vegetable consumption against premenopausal bone loss.
Xie et al. [35]	China	Examine the association between fruit and vegetable consumption with osteoporotic hip fractures risk	Case-control	An inverse association between greater consumption of both fruit and vegetables in men and vegetables in women and risk of osteoporotic hip fractures in elderly Chinese.
Liu et al. [23]	China	Examine the association between fruit intake and bone mineral status	Cross-sectional	A positive association between greater fruit consumption with better bone mineral status among Chinese elderly men and women.
Hammad et al. [7]	Saudi Arabia	Examine the association between lifestyle factors with bone health	Cross-sectional	The association between combined lifestyle factors, including higher soft drink intake, lack of exercise, and limited calcium and vitamin D supplementation with osteopenia and osteoporosis among young Saudi females.
Qiu et al. [12]	China	Examine the association between fruit and vegetable consumption and bone mineral density	Cross-sectional	Association between greater consumption of fruit and vegetables with a higher BMD and lower presence of osteoporosis in middle-aged and elderly Chinese subjects with lower BMD.

Study (Year)	Country	Objective	Type of study	Key findings
Julian et al. [21]	Spain	Examine the association between Mediterranean diet, diet quality with bone mineral content	Cross-sectional	No association between Mediterranean diet score and diet quality index with bone mineral content in Spanish adolescents.
McNaughton et al. [22]	Australia	Examine the association between energy-dense, nutrient-poor dietary pattern, and bone health	Cross-sectional	A direct association between high consumption of legumes, seafood, seeds, nuts, wine, rice, rice dishes, other vegetables, and vegetable dishes and low consumption of bacon and ham with BMD at the lumbar spine and total hip and total BMC in Australian women.
de Jonge et al. [26]	Netherland	Examine the association between dietary patterns and bone mineral density	Cohort	A positive association between health-conscious dietary patterns, including higher consumption of fruit, vegetables, fish, and poultry with BMD independent of anthropometrics.
Kim et al. [39]	Korea	Examine the association between dietary antioxidant capacity and bone mass	Cross-sectional	1. An inverse association between dietary total antioxidant capacity and risk of osteoporosis in postmenopausal women. 2. A positive association between bone mineral content (BMC) and bone mineral density of the femoral neck and lumbar spine in postmenopausal women and BMC of the total femur and lumbar spine in premenopausal women.
Ghadiri et al. [2]	Iran	Examine the association between healthy plant-based diet index with bone mineral density	Case–control	1. A negative association between the last tertile of a healthy plant-based diet with femoral and lumbar BMD abnormality in osteoporotic postmenopausal women. 2. A positive association between the last tertile of an unhealthy plant-based diet with femoral and lumbar BMD abnormality in osteoporotic postmenopausal women.
Ghadiri et al. [4]	Iran	Examine the association between healthy plant-based diet index with bone mineral density	Case-control	1. A negative association between the last tertile of a healthy plant-based diet with femoral and lumbar BMD abnormality in osteoporotic postmenopausal women. 2. A positive association between the last tertile of an unhealthy plant-based diet with femoral and lumbar BMD abnormality in osteoporotic postmenopausal women.

Study (Year)	Country	Objective	Type of study	Key findings
Yoo et al. [34]	Korea	Examine the association between the phytochemical index and bone mineral density	Cohort	Association between the highest quartile of the phytochemical index and lower risk of osteoporosis

Table 1.
Summary of studies.

Author details

Marzieh Mahmoodi[1,2], Fateme Souni[2], Elham Tavassoli Nejad[3], Zainab Shateri[4] and Mehran Nouri[5,6]*

1 Student Research Committee, Shiraz University of Medical Sciences, Shiraz, Iran

2 Department of Clinical Nutrition, School of Nutrition and Food Science, Shiraz University of Medical Sciences, Shiraz, Iran

3 Department of Clinical Nutrition, School of Nutrition and Food Sciences, Isfahan University of Medical Sciences, Isfahan, Iran

4 Department of Nutrition and Biochemistry, School of Medicine, Ilam University of Medical Sciences, Ilam, Iran

5 Mobility Impairment Research Center, Health Research Institute, Babol University of Medical Sciences, Babol, Iran

6 Health Policy Research Center, Institute of Health, Shiraz University of Medical Sciences, Shiraz, Iran

*Address all correspondence to: mehran_nouri71@yahoo.com

References

[1] Varacallo M, et al. Osteopenia. StatPearls; 2018

[2] Ghadiri M et al. The relation between dietary quality and healthy eating index with bone mineral density in osteoporosis: A case-control study. BMC Musculoskeletal Disorders. 2023;**24**(1):584

[3] Grili PPDF et al. Nutrient patterns and risk of osteopenia in postmenopausal women. Nutrients. 2023;**15**(7):1670

[4] Ghadiri M et al. Healthy plant-based diet index as a determinant of bone mineral density in osteoporotic postmenopausal women: A case-control study. Frontiers in Nutrition. 2023;**9**:1083685

[5] Zhou Q et al. Oxidative stress-related biomarkers in postmenopausal osteoporosis: A systematic review and meta-analyses. Disease Markers. 2016;**2016**:7067984

[6] Karaguzel G, Holick MF. Diagnosis and treatment of osteopenia. Reviews in Endocrine and Metabolic Disorders. 2010;**11**(4):237-251

[7] Hammad LF, Benajiba N. Lifestyle factors influencing bone health in young adult women in Saudi Arabia. African Health Sciences. 2017;**17**(2):524-531

[8] Fabiani R, Naldini G, Chiavarini M. Dietary patterns in relation to low bone mineral density and fracture risk: A systematic review and meta-analysis. Advances in Nutrition. 2019;**10**(2):219-236

[9] Jackson MK et al. Dietary inflammatory potential and bone outcomes in Midwestern post-menopausal women. Nutrients. 2023;**15**(19):4277

[10] Rizzoli R. Dairy products, yogurts, and bone health. The American Journal of Clinical Nutrition. 2014;**99**(5):1256S-1262S

[11] van den Heuvel EG, Steijns JM. Dairy products and bone health: How strong is the scientific evidence? Nutrition Research Reviews. 2018;**31**(2):164-178

[12] Qiu R et al. Greater intake of fruit and vegetables is associated with greater bone mineral density and lower osteoporosis risk in middle-aged and elderly adults. PLoS One. 2017;**12**(1):e0168906

[13] Darling AL et al. Dietary protein and bone health: A systematic review and meta-analysis. The American Journal of Clinical Nutrition. 2009;**90**(6):1674-1692

[14] Shams-White MM et al. Animal versus plant protein and adult bone health: A systematic review and meta-analysis from the National Osteoporosis Foundation. PLoS One. 2018;**13**(2):e0192459

[15] Salehi-Sahlabadi A et al. Dietary phytochemical index and the risk of non-alcoholic fatty liver disease: A case-control study among Iranian adults. Complementary Therapies in Medicine. 2022;**71**:102881

[16] Mehranfar S et al. Association of dietary phytochemical index with cardiometabolic risk factors. International Journal for Vitamin and Nutrition Research. 2023;**93**(6):559-576

[17] Bahadoran Z et al. Dietary phytochemical index and the risk of breast cancer: A case control study in a population of Iranian women. Asian

Pacific Journal of Cancer Prevention. 2013;**14**(5):2747-2751

[18] Delshad Aghdam S et al. Dietary phytochemical index associated with cardiovascular risk factor in patients with type 1 diabetes mellitus. BMC Cardiovascular Disorders. 2021;**21**(1):293

[19] Rigi S et al. Dietary phytochemical index in relation to risk of glioma: A case-control study in Iranian adults. Nutrition Journal. 2021;**20**(1):31

[20] Rigi S et al. Dietary phytochemical index in relation to risk of stroke: A case-control study. Nutritional Neuroscience. 2022;**25**(11):2239-2246

[21] Julián C et al. Mediterranean diet, diet quality, and bone mineral content in adolescents: The HELENA study. Osteoporosis International. 2018;**29**(6):1329-1340

[22] McNaughton SA et al. An energy-dense, nutrient-poor dietary pattern is inversely associated with bone health in women. The Journal of Nutrition. 2011;**141**(8):1516-1523

[23] Liu ZM et al. Greater fruit intake was associated with better bone mineral status among Chinese elderly men and women: Results of Hong Kong Mr. Os and Ms. Os studies. Journal of the American Medical Directors Association. 2015;**16**(4):309-315

[24] Macdonald HM et al. Nutritional associations with bone loss during the menopausal transition: Evidence of a beneficial effect of calcium, alcohol, and fruit and vegetable nutrients and of a detrimental effect of fatty acids. The American Journal of Clinical Nutrition. 2004;**79**(1):155-165

[25] Tucker KL et al. Bone mineral density and dietary patterns in older adults: The Framingham osteoporosis study1,2,3. The American Journal of Clinical Nutrition. 2002;**76**(1):245-252

[26] de Jonge EAL et al. Dietary patterns in an elderly population and their relation with bone mineral density: The Rotterdam study. European Journal of Nutrition. 2018;**57**(1):61-73

[27] Brondani JE et al. Fruit and vegetable intake and bones: A systematic review and meta-analysis. PLoS One. 2019;**14**(5):e0217223

[28] Quattrini S et al. The Mediterranean diet in osteoporosis prevention: An insight in a peri- and post-menopausal population. Nutrients. 2021;**13**(2):531

[29] Mirmiran P et al. Association between dietary phytochemical index and 3-year changes in weight, waist circumference and body adiposity index in adults: Tehran lipid and glucose study. Nutrition & Metabolism (London). 2012;**9**(1):108

[30] Vincent HK, Bourguignon CM, Taylor AG. Relationship of the dietary phytochemical index to weight gain, oxidative stress and inflammation in overweight young adults. Journal of Human Nutrition and Dietetics. 2010;**23**(1):20-29

[31] Hanga-Farcaș A et al. Phytochemical compounds involved in the bone regeneration process and their innovative administration: A systematic review. Plants (Basel). 2023;**12**(10):2055

[32] Salvio G et al. The effects of polyphenols on bone metabolism in postmenopausal women: Systematic review and meta-analysis of randomized control trials. Antioxidants (Basel). 2023;**12**(10):1830

[33] Nouri M et al. How do carbohydrate quality indices influence on bone mass

density in postmenopausal women? A case-control study. BMC Women's Health. 2023;**23**(1):42

[34] Yoo H, Park K. Association between phytochemical index and osteoporosis in women: A prospective cohort study in Korea. Nutrients. 2023;**15**(7):1605

[35] Xie HL et al. Greater intake of fruit and vegetables is associated with a lower risk of osteoporotic hip fractures in elderly Chinese: A 1:1 matched case-control study. Osteoporosis International. 2013;**24**(11):2827-2836

[36] Shahriarpour Z et al. Oxidative balance score and risk of osteoporosis among postmenopausal Iranian women. Archives of Osteoporosis. 2021;**16**(1):43

[37] Hernández-Ruiz Á et al. Oxidative balance scores (OBSs) integrating nutrient, food and lifestyle dimensions: Development of the NutrientL-OBS and FoodL-OBS. Antioxidants (Basel). 2022;**11**(2):300

[38] Noruzi Z et al. Association of oxidative balance score with the metabolic syndrome in a sample of Iranian adults. Oxidative Medicine and Cellular Longevity. 2021;**2021**:5593919

[39] Kim D, Han A, Park Y. Association of dietary total antioxidant capacity with bone mass and osteoporosis risk in Korean women: Analysis of the Korea national health and nutrition examination survey 2008-2011. Nutrients. 2021;**13**(4):1149

[40] Domazetovic V et al. Oxidative stress in bone remodeling: Role of antioxidants. Clinical Cases in Mineral and Bone Metabolism. 2017;**14**(2):209-216

[41] Goodman M et al. Hypothesis: Oxidative stress score as a combined measure of pro-oxidant and antioxidant exposures. Annals of Epidemiology. 2007;**17**(5):394-399

[42] Bentyaghoob S et al. Oxidative balance score and dietary phytochemical index can reduce the risk of colorectal cancer in Iranian population. BMC Gastroenterology. 2023;**23**(1):183

[43] Park HM et al. Oxidative balance score inversely associated with the prevalence and incidence of metabolic syndrome: Analysis of two studies of the Korean population. Frontiers in Nutrition. 2023;**10**:1226107

[44] Zhou Q et al. Oxidative stress-related biomarkers in postmenopausal osteoporosis: A systematic review and meta-analyses. Disease Markers. 2016;**2016**:7067984

[45] Muñoz-Garach A, García-Fontana B, Muñoz-Torres M. Nutrients and dietary patterns related to osteoporosis. Nutrients. 2020;**12**(7):1986

[46] Marcucci G et al. Oxidative stress and natural antioxidants in osteoporosis: Novel preventive and therapeutic approaches. Antioxidants (Basel). 2023;**12**(2):209-216

[47] Shen CL et al. Fruits and dietary phytochemicals in bone protection. Nutrition Research. 2012;**32**(12):897-910

Chapter 2

The Role of Antioxidants in Semen Freezing

İbrahim Halil Güngör, Seyfettin Gür and Gaffari Türk

Abstract

In recent years, free radicals and antioxidants are gaining importance in many fields, especially in the field of andrology. With the increase of free radicals in the environment, the deficiency of endogenous antioxidants occurs. Although many molecules are called antioxidants, antioxidants are generally divided into two: endogenous and exogenous antioxidants. Endogenous antioxidants are divided into two groups: enzymatic and non-enzymatic, while exogenous antioxidants are divided into six groups: vitamins, polyphenols, minerals, drugs, chelators, and oxygen collectors. Antioxidants are classified differently according to their mode of action. It is classified as reactive oxygen species by enzyme or directly sweeping, repressing the formation of reactive oxygen species, breaking the radical chain by suppressing metal ions, and repairing the damaged molecule in general, antioxidants are very rich in diversity. As a result, freezing of semen causes an increase in the lipid peroxidation level in the environment due to the lipidic composition of the plasma membrane of the spermatozoa, resulting in membrane damage in the spermatozoa. At this stage, it is important to add different antioxidants to semen extenders to improve the freezing quality of semen.

Keywords: semen, freezing, antioxidants, storability, lipid peroksidasyon

1. Introduction

Most living things have to take in oxygen (molecular oxygen, O_2) to survive. Approximately 90% of the O_2 is used in the electron transport chain (respiratory chain) in the body, and 10% is used in other O_2-requiring reactions. In the electron transport chain, the high oxidizing power of O_2 is converted into the high-energy hydrogen bond of adenosine triphosphate (ATP). O_2 has two unpaired electrons [1, 2]. Molecules that contain one or more unpaired electrons and have the capacity to oxidatively change the structure of biomolecules when they encounter them are defined as "free radicals." Free radicals can be positive, negative, or neutral in terms of electrical charge [3, 4]. Free radicals are simple molecules, atoms, or ions that contain one or more unpaired "electrons" in their atomic or molecular orbitals. In other words, free radicals are molecules in which the number of negatively charged electrons is not equal to the number of positively charged protons. Unpaired electrons in free radicals want to reach a stable state and take electrons from a stable compound, turning this compound into a new free radical. This series of chain reactions initiated by free radicals continues until it is stopped by antioxidants [5, 6]. The greatest

IntechOpen

Radicals	Non-radicals
Superoxide anion (O_2-.)	Hydrogen peroxide (H_2O_2)
Hydroxyl (HO.)	Lipid hydroperoxide (LOOH)
Peroxyl (ROO.)	Hypohalous acid (HOX)
Alkoxyl (RO.)	N-Halogenated amines (R-NH-X)
Semiquinone (HQ.)	Singlet oxygen ($1O_2)_2$
Organic radicals (R.)	Ozone (O_3)
Organic peroxide (RCOO.)	Nitrogen dioxide (NO_2)
Nitric oxide (NO.)	Hypochlorous acid (HOCI)
Hemoprotein bound radicals	Peroxynitrite (ONOO-)

Table 1.
Reactive oxygen species.

damage caused by free radicals is on cell membranes. These pair by stealing electrons from cell membranes and disrupting the cell membrane and cell structure [7, 8]. In terms of terminology, many researchers have offered different definitions to define free oxygen radicals and related (non-radical) species. However, the definition made by Halliwell seems to be more comprehensive. Halliwell states that he prefers to use the definition of reactive oxygen species (ROS) and that with this definition, ROS includes both radical and non-radical O_2 molecules (**Table 1**), accordingly, all oxygen radicals are ROS, but not all ROS are oxygen radicals [2].

Reproductive biotechnology, which is of indispensable importance in the livestock industry, includes some innovations such as artificial insemination, *in vitro* fertilization, gene transfer, and cloning. The most effective and common among these techniques is the artificial insemination technique, which is based on freezing male gamete cells and artificial insemination of female animals [9]. However, this semen freezing process has some known disadvantages. The most important of these disadvantages is the oxidative stress that the cryopreservation process creates on the spermatozoon membrane. In addition to oxidative stress, peroxidation of lipids and proteins in cells and apoptosis are other disadvantages. In addition, although spermatozoon DNA has a tighter (compact) structure than somatic cell DNA, breaks are frequently observed as a result of ROS attacks [10]. In fact, ROS have many functions such as maintaining intracellular balance [11] at physiological levels, phagocytosis, maturation of spermatozoa in the epididymis, capacitation and hyperactivation of spermatozoa in the female genital tract, and acrosome reaction and spermatozoon-oocyte fusion [12, 13].

The most important defense mechanism to eliminate oxidative stress caused by free radicals is antioxidants. Antioxidants are substances that can scavenge free radicals and prevent cell damage. Antioxidants are either produced naturally by the body or taken externally. Both endogenous and exogenous antioxidants act as free radical scavengers [14]. Antioxidants have a protective effect by neutralizing free radicals, which are toxic byproducts of normal cell metabolism [15].

2. Cryoinjuries caused by cryopreservation process

The plasma membrane structure of spermatozoon plays a very important role in fertilization. Therefore, the biochemical structure of the membrane is important in semen freezing processes. Membrane fluidity and elasticity, which are necessary for the normal functions of spermatozoa such as motility, acrosome reaction, and spermatozoon-oocyte fusion, depend entirely on the lipid structure of the membrane. The lipid composition of the spermatozoon membrane differs from that of the somatic cell membrane, with the most characteristic lipid fraction being phospholipids. Phosphatidylcholine, phosphatidylethanolamine, and sphingomyelin are the largest components of phospholipids [16]. Among these phospholipids, polyunsaturated fatty acids (PUFA) are held responsible for the fluidity and elasticity of the spermatozoon membrane [17, 18]. The complete formation of the lipid composition in the sperm plasma membrane occurs during epididymal transport and varies according to species. When the plasma membrane lipids of ejaculated ram, goat, and pig spermatozoa are examined, it is reported that the lipid compositions of spermatozoon of these species undergo significant changes during epididymal maturation.

Cooling and freezing processes during cryopreservation damage spermatozoa due to loss of plasma and acrosome membrane integrity and disruptions in membrane lipids that lead to mitochondrial damage [19]. Although these events vary depending on the animal species, it is known that some species of animals cause motility losses in their sperm. In addition, the cryopreservation process causes an increase in the rates of dead and abnormal spermatozoon and spermatozoon with damaged plasma and acrosomal membranes. Another situation caused by this process is an increase in lipid peroxidation (LPO), which is an indicator of disruption in the oxidant-antioxidant balance, and a decrease in antioxidant activity [20–22]. Another damage is the change in the composition of membrane lipids [23, 24] and DNA fragmentation [25]. It has also been reported that cryopreservation causes an increase in the number of prematurely capacitated spermatozoon [26] and the level of apoptosis [27]. On the other hand, it has been reported that cryopreservation processes cause changes in some gene expressions in pig semen related to epigenetic changes [28] and increases in DNA methylation in stallion semen [29]. Similarly, it has been reported that cryopreservation increases the autophagic protein rates in stallion semen [30]. It is also reported that cryopreservation reduces messenger RNA (mRNA) expressions in human, pig, and bull spermatozoa [31] and miRNA-related protein levels in pig spermatozoa [32]. The resistance of spermatozoa of different species to cold shock damage also varies. This condition is mostly related to the lipid composition in the plasma. Oxidative stress is defined as the imbalance between reactive oxygen species (ROS) and the antioxidant membrane of spermatozoa [33]. It is known that the freezing-thawing process reduces the antioxidant defense capacity and superoxide dismutase (SOD) activity of the spermatozoon cell. Additionally, this process reduces the glutathione level by around 78% compared to fresh semen [34].

3. Oxidative stress

Oxidative stress is defined as the imbalance between ROS and antioxidants. ROS, in addition to its generally known pathological effects, also has physiological effects on cells. The pathological effect is associated with high levels of ROS production, while the physiological effect is associated with low levels of ROS production. The fact that the mammalian spermatozoon membrane is rich in PUFA increases the sensitivity to oxidative stress. While excessive ROS production causes lipid, protein, and carbohydrate peroxidation, DNA damage, and apoptosis in spermatozoa, low ROS production (physiological level) is necessary for the normal functions of spermatozoa such as maturation, capacitation, hyperactivation, acrosome reaction, and sperm-oocyte fusion. Negativities such as disruption in spermatogenesis, decrease in sperm motility and concentration, abnormal spermatozoon ratio, dead-to-live spermatozoon ratio, increase in apoptosis and DNA fragmentation, acrosome, mitochondrial activity and fertilization disorders, and changes in membrane lipid composition are among the spermatozoal damages caused by endogenous and exogenous excessive ROS production [35]. Oxidative damage occurs as a result of the increase of ROS in the cell and their negative impact on cell functions. Although cells can resolve mild oxidative stress on their own, they often use antioxidant systems. However, when intracellular defense systems are inadequate, DNA, protein, carbohydrates, and lipids, which are cellular macromolecules sensitive to oxidant damage, may be damaged [36].

4. Effects of antioxidants on oxidative stress occurring during cryopreservation of semen

Spermatozoa try to protect itself against oxidative stress thanks to the antioxidants found in seminal plasma [37]. The first determined effect of antioxidants was the protection against lipid peroxidation in the cell membrane structure. As a result, antioxidants were initially defined as molecules that inhibit lipid peroxidation. Today, the definition of antioxidants has been expanded to include their protective effects on other target molecules, such as proteins, nucleic acids, and carbohydrates, as well as lipids. Thus, antioxidants are defined as substances that prevent or delay oxidant damage to target molecules, and depending on this definition, the effects of antioxidants can take different forms [38].

The main types of antioxidant effects are:

- Clearing of ROS through enzyme reactions or directly (scavenging effect) [39].
- Preventing the formation of ROS by suppression (quenching effect) [40].
- Preventing radical formation reactions by binding metal ions (chain-breaking effect) [41].
- Repair or cleaning of target molecules after damage (repair effect) [42].

4.1 Classification of antioxidants

Although many substances found in nature are considered antioxidants, antioxidants are basically divided into two groups: endogenous and exogenous antioxidants. Endogenously sourced antioxidants can be classified into two subgroups: enzymatic and nonenzymatic antioxidants [43].

4.1.1 Endogenous antioxidants

Endogenous antioxidants are examined in two parts: enzymatic and nonenzymatic antioxidants (**Table 2**).

Endogenous antioxidants	
Nonenzymatic antioxidants	**Enzymatic antioxidants**
Taurine	Glutathione reductase (GSSG-R)
Trehalose	Glutathione peroxidase (GSH-Px)
Alpha-lipoic acid	Catalase (CAT)
Glutathione (GSH)	Superoxide dismutase (SOD)
Urate	Phospholipid hydroperoxide glutathione
Bilirubin	Peroxidase (PLGPx)
Cysteine	Glutathione-S-transferase
Melatonin	
Ceruloplasmin	
Transferrin	
Ferritin	
Lactoferrin	
Albumin	
Coenzyme Q10	

Table 2.
Classification of endogenous antioxidants.

4.1.1.1 Enzymatic antioxidants

4.1.1.1.1 Superoxide dismutase (SOD)

SOD constitutes the first line of defense against reactive oxygen species [15, 44]. Superoxide dismutase is an enzymatic antioxidant that catalyzes superoxide radical (O_2-) into hydrogen peroxide (H_2O_2) and molecular oxygen (O_2). Hydrogen peroxide is then removed from the environment by catalase (CAT) or glutathione peroxidase (GSH-Px) [45]. There are three forms of SOD in living things. Of these, superoxide dismutase (Cu/Zn SOD) containing copper (Cu) and zinc (Zn) is found in the cytosol, superoxide dismutase (Mn SOD) containing manganese (Mn) is found in the mitochondria, and extracellular superoxide dismutase (EC SOD) is found in extracellular fluids [43, 45].

4.1.1.1.2 Catalase (CAT)

CAT enzyme is a hemoprotein with a tetramer structure. Its function is to break down H_2O_2 into oxygen and water. The organelles with the highest activity are peroxisomes. In addition to the most common role of CAT in the catalytic breakdown of H_2O_2, it is also known to oxidize electron donors such as ethanol or phenols at low concentrations of H_2O_2. Since H_2O_2 production occurs at different levels in different tissues, the actual function of CAT may vary from tissue to tissue and subcellular compartments [46]. While CAT increases the viability of bull spermatozoa in egg yolk extenders, it does not provide this effect in milk-based extenders [47]. In a study conducted to freeze rooster semen, 50, 100, 200, and 300 μg/mL CAT and 50, 100, 200, and 300 U/mL SOD were added to the semen extender.

After freezing-thawing, high motility and viability rates were obtained in extender groups containing 100 μg/mL CAT and 50 U/mL SOD [48]. In a study conducted to freeze pig semen, SOD (150 and 300 U/mL) and CAT (200 and 400 U/mL) were added. It has been reported that both doses of SOD and CAT supplementation after freezing-thawing contribute positively to the motility and kinematic parameters of sperm compared to the control group [49].

4.1.1.1.3 Glutathione reductase (GSSG-R, GR)

GR is a tripeptide widely found in living cells [50]. GSSG-R is a thiol compound that has a protective effect against ROS products formed as a result of lipid peroxidation and does not have a protein structure. This thiol compound plays an important role in the antioxidation and detoxification of exogenous and endogenous compounds in the cell, as well as in maintaining the intracellular redox process. Its role in the antioxidant system is that it has the ability to regenerate water-soluble antioxidants. For this reason, it is used together with antioxidants [50, 51]. The inclusion of GSSG-R in diluents creates a protective effect against oxidative damage by reacting with ROS products in the environment and partially preventing the degradation and dispersion that develops in the sulfur groups in membrane proteins during freezing [52].

4.1.1.1.4 Glutathione peroxidase (GSH-Px)

GSH-Px, an enzymatic antioxidant, is normally found in human, bull, and stallion spermatozoa. GSH-Px protects spermatozoa against both H_2O_2, the most toxic end product of reactive oxygen species, and lipid peroxidation. GSH-Px plays a key role in the defense mechanism against H_2O_2, which is an important factor that causes damage to spermatozoa and DNA [53].

4.1.1.1.5 Phospholipid hydroperoxide glutathione peroxidase (PLGSH-Px)

PLGSH-Px is a monomeric, cytosolic enzyme containing selenium atoms. The membrane reduces phospholipid hydroperoxides to alcohols. When vitamin E, the most important membrane-bound antioxidant, is sufficient, PLGSH-Px protects the membrane against peroxidation [54].

4.1.1.2 Nonenzymatic antioxidants

4.1.1.2.1 Taurine

Taurine, synthesized from cysteine in the organism, is a colorless, water soluble, freely available amino acid that does not participate in the protein structure, and has a molecular weight of 125 g/mol. It is suggested that taurine plays a role in important events such as detoxification, osmoregulation, Ca^{2+} flow regulation, and membrane stabilization [55, 56]. Its protective and supportive effects as an antioxidant, such as maintaining oxidant-antioxidant balance and cell integrity in the body and increasing body resistance, are important [57]. Female genital tract fluid and seminal plasma contain high concentrations of certain amino acids, such as taurine, glutamine, glutamate, and glycine, and these may also play a role in penetration, acrosome reaction, capacitation, and embryo development and preimplantation [58]. Taurine is also found in human semen, sperm acrosome, and rabbit and mouse eggs. It is necessary for maintaining sperm motility and optimal fertilization of hamster oocytes in hamsters and humans [59]. Taurine, a sulfonic amino acid that is also found in epididymal and oviductal fluid, plays an important role in protecting spermatozoa against ROS, especially those formed during freezing-thawing. Taurine and hypotaurine show their effect by reducing motility loss and lipid peroxidation in spermatozoa [60]. In a study conducted to freeze bull semen, it was reported that taurine and hypotaurine added to semen extenders in low amounts positively improved the rate of motile spermatozoon after freezing-thawing [61]. In addition, taurine and hypotaurine support the protective effect of glycerol, which is added to the semen extender as a cryoprotectant [62]. In a study conducted on rabbit semen, it was reported that taurine had a strong effect on the hypochlorite molecule and hydroxyl radical, and a moderate effect on superoxide radicals and hydrogen peroxide molecules [63]. Beneficial effects of taurine, which works as an antioxidant in biological systems, have also been reported, such as reducing the peroxidation of unsaturated membrane lipids, scavenging reactive oxygen species, and stabilizing biomembranes [64]. It has been reported that the addition of 25 and 50 mM taurine to Tris-based ram semen extender increased the percentage of motile spermatozoon after thawing [65]. In another study, it was reported that taurine added to Tris-based ram semen extender had a positive effect on spermatozoon membrane integrity and viability after freezing-thawing [20].

4.1.1.2.2 Trehalose (α-trehalose)

Trehalose is a disaccharide formed by bonding two d-glucose molecules: α_1 and α_1 [66]. Trehalose is produced by various organisms in response to stress situations such as osmotic shock, extreme temperature, and dehydration. However, trehalose can stabilize various biomolecules, including proteins, membranes, cells, and tissues [67]. Although the exact mechanism of action of trehalose is unknown, it is known that it penetrates the spermatozoon plasma membrane and provides surface area by forming hydrogen bonds with the polar head groups of membrane phospholipids during freezing and thawing. It is also suggested that it reduces the release of reactive oxygen species and, along with this release, the consumption of reduced glutathione (GSH) in spermatozoon and sperm plasma. It also provides protection against cold shock by acting as an osmotic buffer between the cell and the environment. In addition to the cryoprotective effect and energy substrate function of trehalose, it contributes to cell dehydration by preventing intracellular ice crystal formation and water flow in the spermatozoa membrane during freezing [68]. Trehalose plays an important role in increasing membrane fluidity before freezing and provides better resistance to spermatozoa against freeze-thaw damage [69]. In a study conducted to freeze Tris-based ram semen, they reported that adding 100 mM trehalose to the semen extender had a positive effect on motility, membrane, and acrosome integrity after freezing-thawing [70, 71].

4.1.1.2.3 Alpha lipoic acid (ALA)

ALA is a coenzyme-like auxiliary molecule in acyl group transfer and the Krebs cycle [72]. ALA, vit C (water soluble) and vit E unlike (fat soluble), can neutralize free radicals in both aqueous and fatty regions of cells, and its oxidized and reduced forms also have antioxidant effects [73]. In addition, ALA increases the amount of intracellular glutathione [74] and coenzyme Q_{10} [75]. ALA, also known as thioctic acid, is a vitamin-like antioxidant and protects the inside and outside of the cell against various free oxygen species [76]. Exogenous addition of this substance increases the level of free ALA, which reduces oxidative stress both *in vivo* and *in vitro* and acts as a potent antioxidant [77]. ALA scavenges hydroxyl radical and hypochlorous acid but is not effective on superoxide and peroxyl radical. Dihydrolipoic acid (DHLA) is a stronger reductant than GSH and prevents lipid peroxidation by scavenging hypochlorous acid and peroxyl and hydroxyl radicals [78, 79]. In a study conducted to freeze semen in red deer, it was reported that ALA increased the motility and viability of spermatozoa after freezing-thawing of semen, while also reducing DNA damage [80].

4.1.1.2.4 Glutathione (GSH)

GSH is synthesized in almost all eukaryotic cells. Therefore, it is found in high concentrations. Glutathione acts as an antioxidant and is also involved in maintaining the redox state of the cell, the functioning of the detoxification system, the synthesis of eicosanoids, the regulation of the cell signaling mechanism, gene expression, and apoptosis [81]. GSH, an important nonenzymatic antioxidant in the body, protects cells against oxidative damage by reacting with free radicals and peroxides [82].

4.1.1.2.5 Melatonin (N-acetyl-5-methoxy-tryptamine)

Melatonin is mainly produced endogenously from the pineal gland and released into the circulation. Melatonin reduces the harmful effects of free radicals. It protects macromolecules from oxidative damage in all intracellular compartments. Melatonin protects both nuclear DNA and mitochondrial DNA, as well as proteins and lipids. Melatonin provides a wide range of protection thanks to its ubiquitous activity as a direct free radical scavenger and an indirect antioxidant. Thus, melatonin scavenges different forms of free radicals and reactive species, including hydroxyl radical, hydrogen peroxide, singlet oxygen, nitric oxide, peroxynitrite anion, and peroxynitric acid. In addition, it stimulates some of the antioxidant enzymes, including SOD, CAT, GSH-Px, and GR. Additionally, experimentally, melatonin increases the intracellular GSH level by stimulating γ-glutamylcysteine synthetase. Melatonin suppresses prooxidative enzymes such as lipoxygenase and nitric oxide synthase. Melatonin strengthens cellular membranes and thus may assist the cell membrane in resisting oxidative damage. Finally, melatonin reduces the production of free radicals and electron leakage by increasing the efficiency of the electron transport system [83].

4.1.1.2.6 Uric acid and bilirubin

Uric acid is thought to be responsible for approximately half of the total antioxidant capacity of the blood. It neutralizes uric acid, hydroxyl, singlet oxygen, superoxide, peroxynitrite anion, peroxynitric acid, and chelates transition metals. It can act as a protector by preventing lipid peroxidation. In addition to being a powerful free radical scavenger, uric acid also acts as chelators of metal ions such as Fe and Cu [84, 85].

Bilirubin is an effective antioxidant and has a chain-breaking effect by affecting peroxyl radicals [86, 87].

4.1.1.2.7 Albumin

Albumin is one of the most important and effective antioxidants in plasma. In healthy adults, approximately 70–80% of cysteine 34 in albumin contains free sulfhydryl groups. The remainder forms a disulfide with various compounds, such as cysteine, homocysteine, or GSH. Thanks to reduced cysteine 34, albumin can scavenge OH. Albumin is the most widely circulated protein in plasma and has a strong antioxidant effect. This effect is due to its properties that trap free oxygen radicals. In general, albumin represents the main plasma protein target of oxidative stress [88]. It was determined that the bovine serum albumin group added *in vitro* to freeze the semen of Kangal dogs provided an increase in motility parameters compared to the control group [89]. In a similar study, it was observed that the albumin-containing group showed higher motility and lower abnormalities than the control group [90].

4.1.1.2.8 Cysteine

Cysteine, one of the thiol-containing amino acids, plays a fundamental role in GSH synthesis. Sulfur-containing amino acids play an important role in determining the flow direction of cysteine between cysteine catabolism and GSH synthesis. Cysteine repairs the gamma glutamate cysteine ligase enzyme, which is a rate-determining enzyme for GSH and taurine synthesis. Therefore, cysteine can be considered a rate-determining amino acid for GSH synthesis. It also plays an important role as an extracellular reducing agent [91, 92]. In a study conducted to freeze ram semen, it was stated that cysteine added to the semen extender showed a protective effect on sperm motility after freezing-thawing [93].

4.1.1.2.9 Coenzyme Q10 (CoQ10, ubiquinone, vitamin Q10, ubidecaquinone, ubidecarenone)

CoQ_{10} is a vitamin-like benzoquinone compound naturally synthesized in the human body. CoQ_{10}, with its high solubility in lipids, is found in almost all cell membranes, as well as in lipoproteins. Additionally, it is a cofactor for at least three mitochondrial enzymes (Complexes I, II, and III), located in the inner membrane of mitochondria, and plays an important role in oxidative phosphorylation. As an antioxidant, CoQ_{10} scavenges free radicals and suppresses lipid and protein peroxidation. Its reduced form, ubiquinol ($CoQH_2$), acts as a lipophilic antioxidant and participates in electron and proton transport in the electron transport system. Ubiquinol donates electrons to neutralize oxidants and shows very strong antioxidant activity. Thus, CoQ_{10} provides effective protection against toxic ROS such as H_2O_2 and $_{O2}$-. CoQ_{10} prevents lipid peroxidation at rates similar to vitamin E. CoQ_{10} works synergistically with α-tocopherol, regenerating its active forms and acting through a similar mechanism as vitamin C [94]. In a study conducted to dilute and freeze antelope semen with various doses of CoQ_{10}, it was observed that 1 μM CoQ_{10} group in frozen-thawed semen increased motility compared to the control group [95].

4.1.1.2.10 Ceruloplasmin and transferrin

Ceruloplasmin is an α_2 serum glycoprotein that carries 95% of the Cu in blood. Transferrin is a carrier protein responsible for transporting Fe^{+3} to cells. Ceruloplasmin binds reversibly to Cu and has an important role in Cu metabolism. Additionally, it acts as ferroxidase and SOD and protects polyunsaturated fatty acids found in erythrocyte membranes from damage by active oxygen species. Although transferrin is found primarily in serum, it is also found in lower concentrations in other body fluids. The main function of transferrin is to transport Fe^{+3} to cells and is also an important growth factor. Ferrous ion (Fe^{+2}) causes oxidative stress by catalyzing the conversion of H_2O_2 to highly toxic OH by the Fenton reaction. Transferrin acts as an antioxidant by reducing the concentration of free ferrous ions [96].

4.1.2 Exogenous antioxidants

The general classification of exogenous antioxidants is given in **Table 3**. According to this classification, exogenous antioxidants are examined under six main headings. These are those used as medicines (**Table 4**), polyphenols (**Table 5**), vitamins (**Table 6**), minerals (**Table 7**), chelators (**Table 8**), and oxygen collectors (**Table 9**).

4.1.2.1 Phenolic acids

Phenolic acids are hydroxy derivatives of aromatic carboxylic acids. They can vary in structure due to differences in the positions and numbers of hydroxyl carbons on the aromatic chains [97, 98]. Studies have found that phenolic acids have strong inhibitory activity against oxidation induced by peroxyl radicals [99].

4.1.2.2 Flavonoids

Flavonoids are diphenylpropanoids with significant antioxidant and chelating properties and are generally found in plants and cannot be synthesized in the living body. Flavonoids are important because they are free radical repellent, regulate enzyme activities, inhibit cell proliferation, have antibiotic and antiallergenic properties, and act as antidiarrheal, ulcer, and anti-inflammatory drugs [100]. Recent studies show that flavonoids prevent oxidative DNA damage by mechanisms other than free radical scavenging. This mechanism can also be achieved by protecting and strengthening antioxidants, such as glutathione-S-transferase, produced by the body. Most flavonoids have the ability to activate glutathione-S-transferase (GST) and act by increasing GST activity to a statistically significant degree. It is thought that GST is effective by detoxifying xenobiotics with mutagenic potential [101]. Anthocyanins, proanthocyanins, and flavanols are among the most important flavonoids in terms of antioxidant capacity.

4.1.2.3 α-Tocopherol (vitamin E)

Vitamin E, which is fat soluble and the most important chain-breaking antioxidant in the cell membrane, increases sperm motility and viability and reduces lipid

Exogenous antioxidants
Used as medicine
Polyphenols
Vitamins
Minerals
Chelates
Those who collect oxygen

Table 3.
Classification of exogenous antioxidants.

Xanthine oxidase inhibitors • Allopurinol • Oxypurinol • Pterine aldehyde • Tungsten
NADPH oxidase inhibitors • Adenosine • Local anesthetics • Calcium channel blockers • Non-steroid anti-inflammatory drugs
Recombinant superoxide dismutase
Trolox-C (Vitamin E analog)
Those that increase endogenous antioxidant activity (Those that increase GPx activity) • Ebselen • Acetylcysteine
Nonenzymatic free radical scavengers • Mannitol • Albumin
Iron redox cycle inhibitors • Desferrioxamine
Neutrophil adhesion inhibitors
Cytokines (TNF and IL-1)
Barbiturates
Iron chelators

Table 4.
Classification of exogenous antioxidants used as drugs.

Phenolic acids
Flavonoids
Butyl-hydroxy-anisole (BHA)
Butyl-hydroxy-toluene (BHT)
Tertiary-butylhydroquinone (TBHQ)

Table 5.
Polyphenols.

peroxidation. Vitamin E has protective properties against oxidative stress and delays cell death. Vitamin E, as a reducing agent, has an inhibitory effect on the functions of ROS within the cell [102]. However, it has been reported that overdose or unnecessary use in cases where the ROS level does not increase may be harmful by blocking its positive effects on capacitation and fertilization [103]. The most active form is α-tocopherol. The most important function of α-tocopherol, a very powerful

α-Tocopherol (Vit E)
β-Carotene (Vit A)
Ascorbic acid (Vit C)
Folic acid (Vit B9)

Table 6.
Vitamins.

Zinc
Selenium
Copper

Table 7.
Minerals.

Citric acid
Amino acids
EDTA
Phosphoric species

Table 8.
Chelates.

Ascorbyl palmitate
Sulfites
Erythorbic acid
Sodium erythorbate

Table 9.
Those who collect oxygen.

antioxidant, is to protect the unsaturated fatty acids in membrane lipids against the attacks of free oxygen species [104]. The antioxidant effect of α-tocopherol is effective at high oxygen concentrations. Therefore, its effects are evident on lipid structures exposed to high oxygen partial pressures, and it is the least toxic among other fat-soluble vitamins [105]. Vitamin E prevents lipid peroxidation by neutralizing alkoxyl radical and lipid peroxyl. These radicals create the proliferation part of lipid peroxidation, causing loss of both sperm motility and membrane fluidity [106]. Vitamin E is the most important lipophilic antioxidant and has a helpful role in maintaining membrane stability due to its presence in the cell membrane. Vitamin E enables the formation of free radicals by sequestering hydrogen atoms from antioxidant molecules rather than unsaturated fatty acids. Thus, they create breaks in the free radical reaction chain, revealing antioxidant radicals known as unreactive [107]. It has been reported that 50 μM vitamin E analog added to semen extenders in rams provided a significant reduction in the lipid peroxidation level after freezing-thawing compared to the control group [108].

4.1.2.4 β-Carotene

β-carotene is the provitamin of vitamin A and is converted to vitamin A when the body is deprived of vitamin A. It is known that β-carotene, like other carotenoids, has fat-soluble antioxidant activity [109]. It has been determined that β-carotene, the metabolic precursor of vitamin A, can suppress singlet oxygen, scavenge superoxide radicals, and play an antioxidant role by interacting directly with peroxy radicals [110].

4.1.2.5 Vitamin C (ascorbic acid)

Vitamin C easily clears reactive oxygen species and reactive nitrogen species, such as superoxide, hydroperoxyl, singlet oxygen, ozone, peroxynitrite, nitrogen dioxide, and hypochlorous acid, thus effectively protecting against oxidative damage. Vitamin C can act as a coantioxidant by regenerating α-tocopherol from α-tocopheroxyl radicals produced by scavenging lipid-soluble radicals [111]. In addition to these antioxidant functions, vitamin C also shows an oxidant behavior by converting Fe^{+3} into Fe^{+2}, which increases lipid peroxidation [112].

4.1.2.6 Folic acid (pteroylglutamic acid, vitamin B9 or vitamin M)

Folic acid is a water-soluble vitamin B member. Folic acid is also necessary for spermatogenesis in men [113]. Folic acid is a very powerful antioxidant that scavenges ROS [114].

4.1.2.7 Zinc

Zinc mineral is a member of the group of exogenous antioxidants. The antioxidant effect of zinc stems from its role in preventing free radical formation and protecting against oxidative stress. Zinc, which is redox stable, replaces redox-reactive metals, such as iron and copper, in critical cellular and extracellular regions. In addition, zinc is included in the structure of superoxide dismutase, an enzyme with an antioxidant effect and metallothioneins, which protect tissues from the harmful effects of free radicals [115]. In a study, a freeze-thaw process was performed on bison semen by diluting it with 0.072, 0.144, 0.288, 0.576, and 1.152 mg/L zinc. It was demonstrated that the group containing 0.288 mg/L zinc after freezing-thawing provided higher progressive motility, viability, and membrane integrity compared with the control group [116].

4.1.2.8 Copper

Copper acts as both an antioxidant and an peroxidant in biological systems. Therefore, it must be taken into the body at optimum levels. In case of insufficient copper intake, the antioxidant balance in the body is disrupted, resulting in increased lipid peroxidation and cell damage. The amount of antioxidant enzymes such as superoxide dismutase, catalase, and glutathione peroxidase decreases in copper deficiency; thus, the body's antioxidant defense system collapses [117].

4.1.2.9 Selenium

Selenium is an integral part of glutathione peroxidase, an antioxidant enzyme that prevents oxidative damage in body tissues [118]. In a study conducted to freeze bison semen, it was determined that groups containing 1 and 2 μg/mL selenium showed higher motility, viability, membrane integrity, and total antioxidant capacity than the control group. Lower DNA damage was observed in these groups [119].

5. Conclusions

The process of freezing sperm has become a major industry, especially in the field of animal husbandry. Thanks to this point where science and technology have come together, semen taken from animals with superior genetic features is subjected to various processes and frozen. After this freezing process, semen can be stored and transported for years in containers containing liquid nitrogen at −196°C. Cryopreservation of semen, which is an indispensable branch of the livestock industry, has some disadvantages. Although these disadvantages vary between species, the most important one is the damage to the spermatozoon membrane as a result of oxidative stress caused by the freezing process. As a result of this damage resulting from lipid peroxidation caused by free radicals, almost all quality parameters, especially sperm motility after freezing and thawing, are negatively affected. The desired condition after cryopreservation is for the semen to be close to the quality parameters before freezing. However, success has not yet been achieved in most species due to lipid peroxidation and physical damage caused by the freezing process. Since this success could not be achieved, antioxidants, which are the most important defense mechanism, began to be used to eliminate oxidative stress caused by free radicals, and scientific studies have been concentrated in this field. Antioxidants have been used because they are substances that can scavenge free radicals and prevent cell damage. For this purpose, endogenous and exogenous antioxidants were added to semen extenders and cryopreservation success was achieved at certain rates.

Conflict of interest

The authors declare no conflict of interest.

Author details

İbrahim Halil Güngör*, Seyfettin Gür and Gaffari Türk
Faculty of Veterinary Medicine, Department of Reproduction and Artificial Insemination, Fırat University, Elazığ, Türkiye

*Address all correspondence to: ihgungor@firat.edu.tr

References

[1] Dröge W. Free radicals in the physiological control of cell function. Physiological Reviews. 2002;**82**(1):47-95. DOI: 10.1152/physrev.00018.2001

[2] Halliwell B. Reactive species and antioxidants. Redox biology is a fundamental theme of aerobic life. Plant Physiology. 2006;**141**(2):312-322. DOI: 10.1104/pp.106.077073

[3] Akpoyraz M, Durak İ. Serbest Radikallerin Biyolojik Etkileri. Ankara Üniversitesi Tıp Fakültesi Mecmuası. 1995;**48**(2):253-262. DOI: 10.1501/Tipfak_0000000463

[4] Delibaş N, Özcankaya R. Serbest Radikaller. SDÜ Tıp Fakültesi Dergisi. 1995;**2**:3

[5] Akkuş İ. Serbest radikaller ve fizyopatolojik etkileri. Konya: Mimoza yayınları, Kuzucular ofset; 1995. pp. 1-151

[6] Yanbeyi S. Aspirin ve antioksidant buthylated hydroxyanisole'ün tavşanlarda eritrosit total katalaz, süperoksit dismutaz ve glutatyon peroksidaz aktiviteleri üzerine etkileri [thesis]. Samsun: Ondokuz Mayıs Üni. Biyoloji Anabilim Dalı; 1999

[7] Kılınç K. Serbest oksijen radikallerinin biyokimyasal etkileri ve metabolizması. Biyokimya Dergisi. 1985;**2**:60-89

[8] Holley AE, Cheeseman KH. Measuring free radical reactions in vivo. British Medical Bulletin. 1993;**49**(3):494-505. DOI: 10.1093/oxfordjournals.bmb.a072626

[9] Nur Z, Ak K. Donmuş spermanın saklanması ve eritilmesi. Uludag University Journal of the Faculty of Veterinary Medicine. 2003;**22**(1-3):97-102

[10] Türk G. Adverse effects of chemotherapeutics on male reproductive system, and protective strategies. Marmara Pharmaceutical Journal. 2013;**17**(2):73-92. DOI: 10.12991/201317376

[11] Kothari S, Thompson A, Agarwal A, du Plessis SS. Free radicals: Their beneficial and detrimental effects on sperm function. Indian Journal of Experimental Biology. 2010;**48**(5):425-435

[12] Ford WCL. Regulation of sperm function by reactive oxygen species. Human Reproduction Update. 2004;**10**(5):387-399. DOI: 10.1093/humupd/dmh034

[13] Pourova J, Kottova M, Voprsalova M, Pour M. Reactive oxygen and nitrogen species in normal physiological processes. Acta Physiologica (Oxf.). 2010;**198**(1):15-35. DOI: 10.1111/j.1748-1716.2009.02039.x

[14] Shinde A, Ganu J, Naik P. Effect of free radicals & antioxidants on oxidative stress: A review. Journal of Dental and Allied Sciences. 2012;**1**(2):6366

[15] Sen S, Chakraborty R, Sridhar C, Reddy YSR, De B. Free radicals, antioxidants, diseases and phytomedicines: Current status and future prospect. International Journal of Pharmaceutical Sciences and Research. 2010;**3**(1):91-100

[16] Mandal R, Badyakar D, Chakrabarty J. Role of membrane lipid fatty acids in sperm cryopreservation. Advances in Andrology, 2014; 9: ID 190542. DOI:10.1155/2014/190542

[17] Ladha S. Lipid heterogeneity and membrane fluidity in a highly polarized cell, the mammalian spermatozoon. Journal of Membrane Biology. 1998;**165**:1-10. DOI: 10.1007/s002329900415

[18] Wassall SR, Stillwell W. Polyunsaturated fatty acid cholesterol interactions: Domain formation in membranes. Biochimica Biophysica Acta-Biomembranes. 2009;**1788**:24-32. DOI: 10.1016/j.bbamem.2008.10.011

[19] Holt WV, Head MF, North RD. Freeze-induced membrane damage in ram spermatozoa is manifested after thawing: Observations with experimental cryomicroscopy. Biology of Reproduction. 1992;**46**:1086-1094. DOI: 10.1095/biolreprod46.6.1086

[20] Bucak MN, Ateşşahin A, Varişli O, Yüce A, Tekin N, Akçay A. The influence of trehalose, taurine, cysteamine and hyaluronan on ram semen microscopic and oxidative stress parameters after freeze-thawing process. Theriogenology. 2007;**67**:1060-1067. DOI: 10.1016/j.theriogenology.2006.12.004

[21] Çoyan K, Başpınar N, Bucak MN, Akalın PP. Effects of cysteine and ergothioneine on post-thawed merino ram sperm and biochemical parameters. Cryobiology. 2011;**63**:1-6. DOI: 10.1016/j.cryobiol.2011.04.001

[22] Demir K, Bakırer Öztürk G, Cirit Ü, Bozkurt HH, Aktaş A, Birler S, et al. Effects of cooling rate on membrane integrity and motility parameters of cryopreserved ram spermatozoa. Kafkas Üniversitesi Veteriner Fakültesi Dergisi. 2015;**21**:61-67. DOI: 10.9775/kvfd.2014.11726

[23] Muller K, Pomorski T, Muller P, Herrmann A. Stability of transbilayer phospholipid asymmetry in viable ram sperm cells after cryotreatment. Journal of Cell Science. 1999;**112**:11-20. DOI: 10.1242/jcs.112.1.11

[24] Towhidi A, Parks JE. Effect of n-3 fatty acids and α-tocopherol on post-thaw parameters and fatty acid composition of bovine sperm. Journal of Assisted Reproduction and Genetics. 2012;**29**:1051-1056. DOI: 10.1007/s10815-012-9834-7

[25] Peris SI, Morrier A, Dufour M, Bailey JL. Cryopreservation of ram semen facilitates sperm DNA damage: Relationship between sperm andrological parameters and the sperm chromatin structure assay. Journal of Andrology. 2004;**25**:224-233. DOI: 10.1002/j.1939-4640.2004.tb02782.x

[26] Robertson L, Watson PF. Calcium transport in diluted or cooled ram semen. Journal of Reproduction and Fertility. 1986;**77**:177-185. DOI: 10.1530/jrf.0.0770177

[27] Nur Z, Zik B, Üstuner B, Sağırkaya H, Özgüden CG. Effects of different cryoprotective agents on ram sperm morphology and DNA integrity. Theriogenology. 2010;**73**:1267-1275. DOI: 10.1016/j.theriogenology.2009.12.007

[28] Zeng C, Peng W, Ding L, He L, Zhang Y, Fang D, et al. A preliminary study on epigenetic changes during boar spermatozoa cryopreservation. Cryobiology. 2014;**69**:119-127. DOI: 10.1016/j.cryobiol.2014.06.003

[29] Aurich C, Schreiner B, Ille N, Alvarenga M, Scarlet D. Cytosine methylation of sperm DNA in horse semen after cryopreservation. Theriogenology. 2016;**86**:1347-1352. DOI: 10.1016/j.theriogenology.2016.04.077

[30] Gallardo Bolanos JM, Miro Morn AM, Balao da Silva CM, Plaza

Davila M, Martin Munoz P, Aparicio IM, et al. During cooled storage the extender influences processed autophagy marker light chain 3 (LC3B) of stallion spermatozoa. Animal Reproduction Science. 2014;**145**(1-2):40-46. DOI: 10.1016/j.anireprosci.2014.01.001

[31] Yeste M. Sperm cryopreservation update: Cryodamage, markers, and factors affecting the sperm freezability in pigs. Theriogenology. 2016;**85**:47-64. DOI: 10.1016/j.theriogenology.2015.09.047

[32] Zhang Y, Zeng CJ, He L, Ding L, Tang KY, Peng WP. Selection of endogenous reference microRNA genes for quantitative reverse transcription polymerase chain reaction studies of boar spermatozoa cryopreservation. Theriogenology. 2015;**83**:634-641. DOI: 10.1016/j.theriogenology.2014.10.027

[33] White IG. Lipids and calcium uptake of sperm in relation to cold shock and preservation: A review. Reproduction Fertility and Development. 1993;**5**:639-658. DOI: 10.1071/RD9930639

[34] Lasso JL, Noiles EE, Alvarez JG, Storey BT. Mechanism of superoxide dismutase loss from human sperm cells during cryopreservation. Journal of Andrology. 1994;**15**(3):225-265. DOI: 10.1002/j.1939-4640.1994.tb00444.x

[35] Türk G. Reaktif oksijen türlerinin spermatozoon fonksiyonları üzerindeki fizyolojik ve patolojik etkileri. Turkiye Klinikleri Journal of Reproduction and Artificial Insemination-Special Topics. 2015;**1**:26-34

[36] Tekcan M. Oksidatif stres-antioksidan sistemler ve testis. Androloji Bülteni. 2009;**37**:131-136

[37] Kim JG, Parthasarathy S. Oxidation and the spermatozoa. In: Seminars in Reproductive Endocrinology. Vol. 16 (4). Stuttgart, Germany: Thieme Medical Publishers, Inc.; 1998. pp. 235-339. DOI: 10.1055/s-2007-1016283

[38] Rangan U, Bulkley GB. Prospects for treatment of free radical-mediated tissue injury. British Medical Bulletin. 1993;**49**(3):700-718. DOI: 10.1093/oxfordjournals.bmb.a072641

[39] Aydilek N, Aksakal M. Testosteronun tavşanlarda karaciğer antioksidan sistemi üzerine etkisi. Yüzüncü Yıl Üniversitesi Veteriner Fakültesi Dergisi. 2003;**14**(2):22-25

[40] Baskin SI, Salem H. Oxidants, Antioxidants and Free Radicals. Washington DC: Taylor and Francis. CRC Press; 1997. pp. 1-16

[41] Rossell JB. Measurement of rancidity. In: Hllin JC, Hamilton RJ, editors. Rancidity in Foods. London: Blackie Academic; 1983. pp. 26-28

[42] Cuvelier ME, Richard H, Berset C. Comparison of the antioxidative activity of some acidphenols structure-activity relationship. Bioscience, Biotechnology and Biochemistry. 1992;**56**(2):32425. DOI: 10.1271/bbb.56.324

[43] Sen S, Chakraborty R. The role of antioxidants in human health, chapter 1. In: Oxidative Stress: Diagnostics, Prevention and Therapy. Washington, DC: American Chemical Society; 2011. pp. 1-37. DOI: 10.1021/bk-2011-1083.ch001

[44] Folkes LK, Candeias LP, Wardman P. Kinetics and mechanisms of hypochlorous acid reactions. Archives of Biochemistry and Biophysics. 1995;**323**(1):120-126. DOI: 10.1006/abbi.1995.0017

[45] Young IS, Woodside JV. Antioxidants in health and disease. Journal of Clinical Pathology. 2001;**54**(3):176-186. DOI: 10.1136/jcp.54.3.176

[46] Percy ME. Catalase: An old enzyme with a new role. The International Journal of Biochemistry & Cell Biology. 1984;**62**:1006-1014. DOI: 10.1139/o84-129

[47] Foote RH. Survival of bull sperm in milk and yolk extenders with added catalase. Journal of Dairy Science. 1962;**45**:907-910. DOI: 10.3168/jds.S0022-0302(62)89520-4

[48] Amini MR, Kohram H, Zare-Shahaneh A, Zhandi M, Sharideh H, Nabi MM. The effects of different levels of catalase and superoxide dismutase in modified Beltsville extender on rooster post-thawed sperm quality. Cryobiology. 2015;**70**(3):226-232. DOI: 10.1016/j.cryobiol.2015.03.001

[49] Roca J, Rodriguez MJ, Gil MA, Carvajal G, Garcia EM, Cuello C, et al. Survival and *in vitro* fertility of boar spermatozoa frozen in the presence of superoxide dismutase and/or catalase. Journal of Andrology. 2005;**26**:15-24. DOI: 10.1002/j.1939-4640.2005.tb02867.x

[50] Irvıne DS. Gluthatione as a treatment for male infertility. Reviews of Reproduction. 1996;**1**:6-12

[51] Baumber J, Ball BA, Gravence CG, Medina V, Daviesmorel MCG. The effect of reactive oxygen species on equine sperm motility, viability, acrosomal integrity, mitochondrial membrane potential and membrane lipit peroxidation. Journal of Andrology. 2000;**21**:895-902. DOI: 10.1002/j.1939-4640.2000.tb03420.x

[52] Chatterjee S, De Lamirande E, Gagnon C. Cryopreservation alters membrane sulfhydryl status of bull spermatozoa: Protection by oxidized glutathione. Molecular Reproduction and Development. 2001;**60**:498-506. DOI: 10.1002/mrd.1115

[53] Drevet JR. The antioxidant glutathione peroxidase family and sperm: A complex story. Molecular and Cellular Endocrinology. 2006;**250**:70-79. DOI: 10.1016/j.mce.2005.12.027

[54] Mayers PA. Pentoz Fosfat Yolu Ve Heksoz Metabolizmasının Diğer Yolları. In: Menteş G, Ersöz B, editors. Harper'ın Biyokimyası. İstanbul: Barış Kitapevi; 1993. pp. 237-248

[55] Jacobsen JG, Smith LH. Biochemistry and physiology of taurine derivatives. Physiological Reviews. 1968;**48**(2):424-511. DOI: 10.1152/physrev.1968.48.2.424

[56] Kendler BS. Taurine: An overview of its role in preventive medicine. Preventive Medicine. 1989;**18**(1):79-100. DOI: 10.1016/0091-7435(89)90056-X

[57] Parcell S. Sulfur in human nutrition and applications in medicine alternative medicine review. Alternative Medicine Review. 2002;**7**(1):22-24

[58] Barna J, Ashızawa K, Boldizsar H, Inoue M. Effects of taurine on the motility and intracellular free Ca^{2+} concentration of fowl spermatozoa *in vitro*. Reproduction. 1998;**114**(2):225-229. DOI: 10.1530/jrf.0.1140225

[59] Devreker F, Van Den Berg M, Biramane J, Winston RML, Englert Y, Hardy K. Effects of taurine on human embryo development *in vitro*. Human Reproduction. 1999;**14**(9):2350-2356. DOI: 10.1093/humrep/14.9.2350

[60] Saleh RA, Agarwal A. Oxidative stress and male infertility: From research bench to clinical practise.

Journal of Andrology. 2002;**23**(6):737-752. DOI: 10.1002/j.1939-4640.2002.tb02324.x

[61] Chen Y, Foote RH, Brockett CC. Effect of sucrose, Trehalose, Hypotaurine, taurine and blood serum on survival of frozen bull sperm. Crybiology. 1993;**30**:423-431. DOI: 10.1006/cryo.1993.1042

[62] Kundu CN, Das K, Majumder GC. Effect of amino acids on goat cauda epididymal sperm cryopreservation using a chemically defined model system. Cryobiology. 2001;**41**;21-27. DOI: 10.1006/cryo.2001.2296

[63] Aruoma OI, Hallivell B, Hoey BM, Buttler J. The antioxidant action of taurine, hypotaurine and their metabolic precursors. Biochemical Journal. 1998;**256**:251-255. DOI: 10.1042/bj2560251

[64] Kılıç N, Yıldırım Z. Effects of taurine and age on liver antioxidant status and protein oxidation. Turkish Journal of Biochemistry. 2008;**33**(4):169-174

[65] Sanchez-Partida LG, Setchel BP, Maxwell WMC. Epididymal compounds and antioxidants in diluents for the frozen storage of ram spermatozoa. Reproduction Fertility of Development. 1997;**9**(7):689-696. DOI: 10.1071/r97045

[66] Patist A, Zoerb H. Preservation mechanism of trehalose in food and biosystems. Colloids and Surfaces B: Biointerfaces. 2005;**40**(2):107-113. DOI: 10.1016/j.colsurfb.2004.05.003

[67] Hounsa C, Brandt E, Thevelein J, Hohmann S, Prior B. Role of trehalose in survival of saccharomyces cerevisiae under osmotic stres. Microbiology. 1998;**144**:671-680. DOI: 10.1099/00221287-144-3-671

[68] Aisen EG, Medina VH, Venturino A. Cryopreservation and postthawed fertility of ram semen frozen in different trehalose concentrations. Theriogenology. 2002;**57**:1801-1808. DOI: 10.1016/s0093-691x(02)00653-2

[69] Aboagla EME, Terada T. Trehalose-enhanced fluidity of the goat sperm membrane and its protection during freezing. Biology of Reproduction. 2003;**69**(4):1245-1250. DOI: 10.1095/biolreprod.103.017889

[70] Tonietto RA, Goularte KL, Gastal GDA, Schiavon RS, Deschamps JC, Lucia T. Cryoprotectant effect of trehalose and low-density lipoprotein in extenders for frozen ram semen. Small Ruminant Research. 2010;**93**(2-3):206-209. DOI: 10.1016/j.smallrumres.2010.05.003

[71] Matsuoka T, Imai H, Kohno H, Fukui Y. Effects of bovine serum albumin and trehalose in semen diluents for improvement of froze-thawed ram spermatozoa. Journal of Reproduction and Development. 2006;**52**(5):675-683. DOI: 10.1262/jrd.18033

[72] Reed LJ, Debusk BG, Gunsalus IC, Hornberger CS. Crystalline alpha-lipoic acid: A catalytic agent associated with pyruvate dehydrogenase. Science. 1951;**27**:93-94. DOI: 10.1126/science.114.2952.93

[73] Packer L, Witt EH, Tritschler HJ. Alpha-lipoic acid as a biological antioxidant. Free Radical Biology and Medicine. 1995;**19**:227-250. DOI: 10.1016/0891-5849(95)00017-r

[74] Busse E, Zimmer G, Schopohl B, Kornhuber B. Influence of alpha-lipoic acid on intracellular glutathione in vitro and in vivo. Arzneimittel-Forschung. 1992;**42**(6):829-831

[75] Kağan V, Serbinova E, Packer L. Antioxidant effects of ubiquinones in microsomes and mitochondria are mediated by tocopherol recycling. Biochemical and Biophysical Research Communications. 1990;**169**(3):851-857. DOI: 10.1016/0006-291x(90)91971-t

[76] Suzuki YJ, Tsuchıya M, Packer L. Thioctic acid and dihydrolipoic acid are novel antioxidants which interact with reactive oxygen species. Free Radical Research Communications. 1991;**15**(5):255-263. DOI: 10.3109/10715769109105221

[77] Selvakumar E, Prahalathan C, Mythili Y, Varalakshmi P. Beneficial effects of dl-α-lipoic acid on cyclophosphamide-induced oxidative stress in mitochondrial fractions of rat testis. Chemico-Biological Interactions. 2005;**152**(1):59-66. DOI: 10.1016/j.cbi.2005.01.009

[78] Coleman MD, Eason RC, Bailey CJ. The therapeutic use of lipoic acid in diabetes: A current perspective. Environmental Toxicology and Pharmacology. 2001;**10**(4):167-172. DOI: 10.1016/s1382-6689(01)00080-1

[79] Roy S, Packer L. Redox regulation of cell functions by alpha-lipoate: Biochemical and molecular aspects. BioFactors. 1998;**7**(3):263-267

[80] Ibrahim SF, Osman K, Das S, Othman AM, Majid NA, Rahman MPA. A study of the antioxidant effect of alpha lipoic acids on sperm quality. Clinics. 2008;**63**:545-550. DOI: 10.1590/s1807-59322008000400022

[81] Townsend DM, Tew KD, Tapiero H. The importance of glutathione in human disease. Biomedicine & Pharmacotherapy. 2003;**57**(3-4):145155. DOI: 10.1016/s0753-3322(03)00043-x

[82] Glutathione BE. Red Cell Metabolism a Manual of Biochemical Methods. 2nd. ed. New York: Grune and Stratton; 1975

[83] Reiter RJ, Acuna-Castroviejo D, DunXian T, Burkhardt S. Free radical-mediated molecular damage. Mechanisms for the protective actions of melatonin in the central nervous system. Annals of the New York Academy of Sciences. 2006;**939**(1):200-215. DOI: 10.1111/j.17496632.2001.tb03627.x

[84] Kumar AN, Aruna P, Naidu JN, Kumar R, Srivastava AK. Review of concepts and controversies of uric acid as antioxidant and pro-oxidant. Archives Medical Review Journal. 2015;**24**(1):19-40. DOI: 10.17827/aktd.53469

[85] Waring WS. Uric acid: An important antioxidant in acute ischaemic stroke. QJM. 2002;**95**(10):691-693. DOI: 10.1093/qjmed/95.10.691

[86] Gutteridge JMC. Lipid peroxidation and antioxidants as biomarkers of tissue damage. Clinical Chemistry. 1995;**41**(12):1819-1828. DOI: 10.1093/clinchem/41.12.1819

[87] Burtis CA, Ashwood ER. Klinik Kimyada Temel İlkeler. 5th ed. Ankara: Palme Yayınları; 2005. pp. 450-452

[88] Roche M, Rondeau P, Singh NR, Tarnus E, Bourdon E. The antioxidant properties of serum albümin. FEBS Letters. 2008;**582**(13):1783-1787. DOI: 10.1016/j.febslet.2008.04.057

[89] Koçyiğit A, Uslu BA. Kangal Irkı Köpeklerin Spermalarının Kriyoprezervasyonunda Sığır Serum Albüminin Etkisi. Cumhuriyet Üniversitesi Sağlık Bilimleri Enstitüsü Dergisi. 2016;**1**(2):47-52

[90] Tosun H, Uysal O. Köpek spermasının dondurulmasında farklı sulandırıcıların ve bireylerin etkisi.

Ankara Üniversitesi Veteriner Fakültesi Dergisi. 2007;**54**(1):23-28

[91] Kim YG, Kim SG, Kwon JW, Park OJ, Kim SK, Kim YC. Effect of cysteine on amino acid concentration and transsulfuration enzyme activities in rat liver with protein-calorie malnutrition. Life Sciences. 2003;**72**(10):1171-1181. DOI: 10.1016/S0024-3205(02)02366-4

[92] Puka-Sundvall M, Sandberg M, Hagberg H. Injury after neonatal hypoxia-ischemia in rats: A role of cysteine? Brain Research. 1998;**797**(2):328-332. DOI: 10.1016/s0006-8993(98)00415-6

[93] Uysal O, Bucak MN. Effects of oxidized glutathione, bovine serumalbumin, cysteine and lycopene on the quality of frozen-thawed ram semen. Acta Veterinaria Brno. 2007;**76**:383-390. DOI: 10.2754/avb200776030383

[94] Gürkan AS, Bozdağ-Dündar O. Coenzyme Q10. Journal of Faculty of Pharmacy of Ankara. 2005;**34**(2):129-154

[95] Yousefian I, Emamverdi M, Karamzadeh-DehaghaniA,Sabzian-MeleiR, Zhandi M, Zare-Shahneh A. Attenuation of cryopreservation-induced oxidative stress by antioxidant: Impact of coenzyme Q10 on the quality of post-thawed buck spermatozoa. Cryobiology. 2018;**81**:88-93. DOI: 10.1016/j.cryobiol.2018.02.005

[96] Chauhan A, Chauhan V, Brown WT, Cohen I. Oxidative stress in autism: Increased lipid peroxidation and reduced serum levels of ceruloplasmin and transferrin - the antioxidant proteins. Life Science. 2004;**75**:2539-2549. DOI: 10.1016/j.lfs.2004.04.038

[97] Andreasen MF, Kroon PA, Williamson G, Garcia-Conesa MT. Intestinal release and uptake of phenolic antioxidant diferulic acids. Free Radical Biology and Medicine. 2001;**31**:304-314. DOI: 10.1016/s0891-5849(01)00585-8

[98] Yu L, Haley S, Perret J, Harris M, Wilson J, Qian M. Free radical scavenging properties of wheat extracts. Journal of Agricultural and Food Chemistry. 2002;**50**:1619-1624. DOI: 10.1021/jf010964p

[99] Neo YP, Ariffin A, Tan CP, Tan YA. Phenolic acid analysis and antioxidant activity assessment of oil palm (E. Guineensis) fruit extracts. Food Chemistry. 2010;**122**:353-359. DOI: 10.1016/j.foodchem.2010.02.046

[100] Çapanoğlu E, Boyacıoğlu D. Meyve ve Sebzelerin Flavonoid İçeriği Üzerine İşlemenin Etkisi. Akademik Gıda. 2009;**7**:41-46

[101] Coşkun T. Fonksiyonel besinlerin sağlığımız üzerine etkileri. Çocuk Sağlığı ve Hastalıkları Dergisi. 2005;**48**:69-84

[102] Akiyama M. In vivo scavenging effects of ethylcysteine on reactive oxygen species in human semen. Nippon Hinyokika Gakkai Zasshi. The Japanese Journal of Urology. 1990;**90**(3):421-428. DOI: 10.5980/jpnjurol1989.90.421

[103] Donnelly ET, Mcclure N, Lewis SE. Antioxidant supplementation in vitro does not improve human sperm motility. Fertility and Sterility. 1999;**72**(3):484-495. DOI: 10.1016/s0015-0282(99)00267-8

[104] Rice D, Kennedy S. Vitamin E: Function and effects of deficiency. British Veterinary Journal. 1988;**144**(5):482-496. DOI: 10.1016/0007-1935(88)90089-9

[105] Deichsel K, Palm F, Koblischke P, Budik S, Aurich C. Effect of a dietary antioxidant supplementation on semen quality in pony stallions. Theriogenology. 2008;**69**:940-945. DOI: 10.1016/j.theriogenology.2008.01.007

[106] Agarwal A, Saleh RA, Bedaiwy MA. Role of reactive oxygen species in the pathophysiology of human reproduction. Fertility and Sterility. 2003;**79**:829-843. DOI: 10.1016/s0015-0282(02)04948-8

[107] Özden S, Catalgol B, Gencinci S, Arda P, Bolkent S, Alpertunga B. Methiocarb-induced oxidative damage following subacute exposure and the protective effects of vitamin E and taurine in rats. Food and Chemical Toxicology. 2009;**47**:1676-1684. DOI: 10.1016/j.fct.2009.04.018

[108] Silva MS, Bicudo SD, Sicherle CC, Rodello L, Gallego IC. Lipit peroxidation and generation of hydrogen peroxidein frozen-thawed ram semen cryopreserved in extenders with antioxidants. Animal Reproduction Science. 2010;**122**:118-123. DOI: 10.1016/j.anireprosci.2010.08.004

[109] Woodall AA, Britton G, Jackson MJ. Carotenoids and protection of phospholipids in solution or in liposomes against oxidation by peroxyl radicals: Relationship between carotenoid structure and protective ability. Biochimica et Biophysica Acta (BBA)-General Subjects. 1997;**1336**(3):575-586. DOI: 10.1016/s0304-4165(97)00007-x

[110] Frei B. Reactive oxygen species and antioxidant vitamins mechanisms of action. The American Journal of Medicine. 1994;**97**(3):5-13. DOI: 10.1016/0002-9343(94)90292-5

[111] Carr AC, Frei B. Toward a new recommended dietary allowance for vitamin C based on antioxidant and health effects in humans. The American Journal of Clinical Nutrition. 1999;**69**(6):1086-1107. DOI: 10.1093/ajcn/69.6.1086

[112] Dündar Y, Aslan R. Oksidan-Antioksidan Denge ve Korunmasında Vitaminlerin Rolü. Hayvancılık Araştırma Dergisi. 1999;**9**(12):32-39

[113] Hussein HK, Elnaggar MH, Al-Zahrani NK. Antioxidant role of folic acid against reproductive toxicity of cyhalothrin in male mice. Global Advanced Research Journal of Environmental Science and Toxicology. 2012;**1**(4):66-71

[114] Ebaid H, Bashandy SAE, Alhazza IM, Rady A, El-Shehry S. Folic acid and melatonin ameliorate carbon tetrachlorideinduced hepatic injury, oxidative stress and inflammation in rats. Nutrition & Metabolism. 2013;**10**(20):1-10. DOI: 10.1186/1743-7075-10-20

[115] Belgemen T, Akar N. Çinkonun Yaşamsal Fonksiyonları ve Çinko Metabolizması ile İlişkili Genler. Ankara Üniversitesi Tıp Fakültesi Mecmuası. 2004;**57**:161-166

[116] Dorostkar K, Shoushtari SMA, Khaki A. Effects of in vitro zinc sulphate additive to the semen extender on water buffalo (bubalusbubalis) spermatozoa before and after freezing. International Journal of Fertility & Sterility. 2014;**8**(3):325

[117] Velioğlu S. Doğal Antioksidanların İnsan Sağlığına Etkileri. Gıda. 2000;**25**(3):167-176

[118] Lobanov AV, Hatfield DL, Gladyshev VN. Reduced reliance on the trace element selenium during evolution of mammals. Genome Biology. 2008;**9**:1-11. DOI: 10.1186/gb-2008-9-3-r62

[119] Dorostkar K, Alavi-Shoushtari SM, Mokarizadeh A. Effects of in vitro selenium addition to the semen extender on the spermatozoa characteristics before and after freezing in water buffaloes (Bubalus bubalis). In: Veterinary Research Forum. Vol. 3 (4). Urmia, Iran: Faculty of Veterinary Medicine, Urmia University; 2012. p. 263

Chapter 3

A Molecular Exploration of Fatty Acid Oxidation and Multiple Acyl Coenzyme A Deficiency

Naini Nishita Rao

Abstract

Fats are the human body's most energy-dense macronutrients with oxidoreductase reactions being fundamental to yielding adenosine triphosphate (ATP), the energy currency of our body. In periods of catabolic stress, fats become vital to energy generation. Multiple acyl-coenzyme A dehydrogenase deficiency (MADD), is a rare fatty acid oxidation disorder that impairs the metabolism of fats and can lead to life-threatening complications. To understand the pathophysiology of MADD, it is fundamental to understand the various oxidoreductase reactions involved in fatty acid metabolism at the molecular level. This chapter will delve into the normal biochemistry of fatty acid metabolism and how the reduced energy shuttles nicotinamide adenine dinucleotide (NADH) and flavin adenine dinucleotide ($FADH_2$) transfer their electrons to the mitochondrial electron transport chain (ETC) to generate ATP. The chapter will go on to explore how an error in these reactions can lead to the biochemical abnormalities and symptomatology seen in MADD.

Keywords: fatty acid oxidation, multiple acyl co-a dehydrogenase deficiency, flavin adenine dinucleotide (FAD), nicotinamide adenine dinucleotide (NAD), electron transport chain (ETC)

1. Introduction

Oxidoreductase reactions are pivotal to the metabolic process of converting digested foods to fuel for the human body with the three major macronutrient groups consisting of carbohydrates, fats and proteins. Glycogen is the major component of endogenous carbohydrates and is stored primarily in the skeletal muscles and liver. Glycogen is essentially polymerised glucose molecules and is the primary way in which the human body stores glucose [1, 2]. During periods of catabolic stress, the body's preferred choice of energy substrate is glucose as glycolysis requires less oxygen and can occur in the cytoplasm of cells making it a more efficient fuel source [1–4]. Under a steady physiological state, glucose is the primary source of energy for the brain and blood glucose levels are tightly regulated by the endocrine system to ensure there is a steady and constant supply. It is estimated that during periods of fasting, glycogen stores can be depleted in 24 hours [3–4]. Once exhausted, the body reverts to stored fats as the source of energy [3].

The breakdown of fatty acids in the human body produces acetyl-CoA, a vital molecule for several biochemical processes including oxidative phosphorylation and gluconeogenesis. The vast majority of fatty acids in humans undergo metabolism by β-oxidation mainly in the mitochondria but this also occurs to a minor degree in the peroxisomes, and alpha-oxidation endoplasmic reticulum [5–8]. The denotations β and α simply refer to the position of the carbon atom in the fatty acid chain that is being oxidised in the redox reaction. The aim of this chapter will be to explore at the molecular level, the pathways and processes of how mitochondrial β-oxidation of fatty acids links to the electron transport chain (ETC) that generates the life-sustaining adenosine triphosphate (ATP) molecules. This will serve as the foundation to understand how a fatty acid oxidation disorder named multiple acyl-coenzyme A deficiency (MADD), can lead to downstream errors in biochemical processes resulting in accumulation of substances such as ammonia, glucose and organic fatty acids, as well as how this can manifest in patient symptomatology.

2. The journey of fats: consumption to oxidative phosphorylation

Although carbohydrates are the preferred source of energy substrate for the body, fat is by far the most abundantly stored endogenous source of energy and becomes vital in times of starvation and prolonged metabolic stress [2–4]. As glycogen is to glucose, triacylglycerols, more commonly known as triglycerides, are the storage form of fats. Triacylglycerols are composed of glycerol backbone (CH2OH-CH2OH-CH2OH) that has an ester link via the hydroxyl -OH group to 3 fatty acids (CH3-(CH2)n-COOH) (see **Figure 1**) [8, 9].

Triacylglycerols make up most digested fats we consume. Digested fats are absorbed in the form of a mixed micelle from the gut lumen via diffusion and enter the bloodstream and lymphatic lacteals as chylomicrons [8, 10]. Once the lacteals empty the chylomicrons into the bloodstream via the thoracic duct, various lipases can break down triacylglycerols to form free fatty acids. These free fatty acids are then

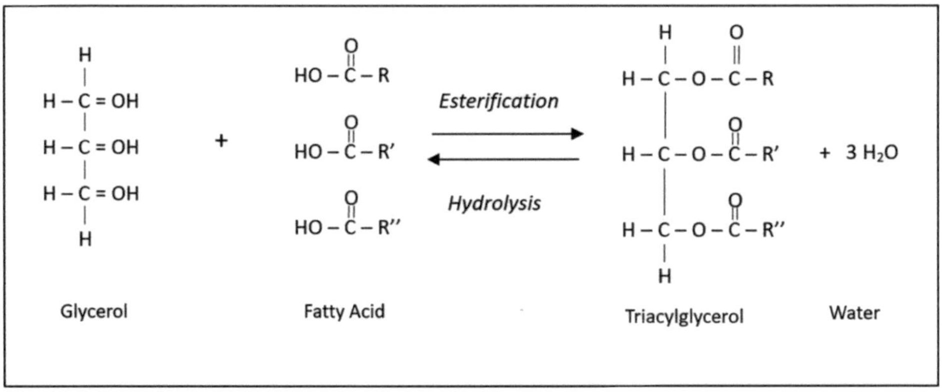

Figure 1.
Esterifaction of glycerol and fatty acids to form triaclglycerol and water. R, alkyl group of unspecified length; R', a second alkyl group of a differnt unspecified length; R', a third alkyl group of a differnt unspecified length. The reverse process is a hydrolysis reaction.

taken up directly by adipocytes and other cells and are reformed into triacylglycerols for storage [10, 11]. The length of the fatty acids is determined by the number of the alkyl groups (n) and are typically classified as short-chain (n < 6), medium-chain (n = 6–13), long-chain (n = 14–17), and very long-chain (n > 18) fatty acids.

During times of increased energy demand, stored triacylglycerols undergo a sequential hydrolysis reaction to re-create free fatty acids, from which the overwhelming majority of energy is derived [5, 10, 11]. The free fatty acids in the serum are taken up by target cells and now must make their way to the inner mitochondrial membrane to undergo β-oxidation [5, 11]. Prior to this, fatty acids are first activated by combining them with a coenzyme A (CoA) molecule via a thioester bond. This is an ATP-dependent reaction catalysed by the enzyme acyl-CoA synthetase found on the outer mitochondrial membrane in a 2-step process requiring 2 high energy phosphate bonds. The end result is the formation of fatty acyl-CoA, adenosine monophosphate (AMP) and a pyrophosphate group [10, 11]. The pyrophosphate group goes on to become hydrolysed by a pyrophosphatase into two phosphate groups – this hydrolysis reaction has a very negative delta G and is the forward-driving thermodynamic force favouring the formation of the activated fatty acyl-CoA [12]. It is important to note that the specific synthetase required to activate very long-chain fatty acids can only be found in peroxisomes which is where they are oxidised as opposed to the mitochondria [6, 10].

While short and medium-chain fatty acids are small enough to traverse the inner mitochondrial membrane without first being activated and then undergo conversion to fatty acyl-CoA in the mitochondria matrix, long-chain fatty acyl-CoA must be conjugated with carnitine first [2, 5, 7, 11, 13]. This is accomplished by the enzyme carnitine-palmitoyl transferase 1 (CPT 1) to produce fatty acylcarnitine and coenzyme A [11, 13]. The fatty acyl carnitine located in the inter mitochondrial membrane acylcarnitine translocase located on the inner mitochondrial membrane transports fatty acylcarnitine to the mitochondrial matrix in exchange for carnitine from the mitochondrial matrix to the inter mitochondrial membrane space in a 1:1 ratio [11, 13]. Once in the matrix, carnitine-palmitoyl transferase 2 also located on the inner part of the inner mitochondrial membrane then uses CoA to displace the fatty acyl carnitine bond to re-form fatty acyl-CoA and carnitine. This carnitine molecule is then recycled back to the outer membrane via acylcarnitine translocase [11, 13]. Now that the activated fatty acyl-CoA is in the mitochondrial matrix, it can undergo β-oxidation via a 4-step process.

Step 1

The first step is an oxidation reaction catalysed by an acyl-CoA dehydrogenase enzyme, that as the name suggests, dehydrogenates carbon 2 and carbon 3 of the fatty acyl Co-A to create a trans-double bond [5, 9, 11]. Flavin adenine dinucleotide (FAD) acts as the acceptor of the 2 hydrogens and becomes reduced to FADH2 creating the end product, a β-enoyl-CoA (see **Figure 2**). There are three types of acyl-CoA dehydrogenases, these isoenzymes are responsible for catalysing short-, medium- and long-chain fatty acyl-CoA molecules respectively [5, 9, 11]. This FADH2 molecule donates the reductive potential to the electron transport chain to eventually produce ATP and becomes recycled into FAD which can help catalyse further dehydrogenation reactions [5, 11, 14].

Step 2

The second step is a hydration reaction catalysed by an enoyl hydratase enzyme. Water is added to saturate the trans-double bond formed in step one to create a β-hydroxyacyl-CoA molecule in the L-isomer configuration (see **Figure 3**) [5, 11, 14].

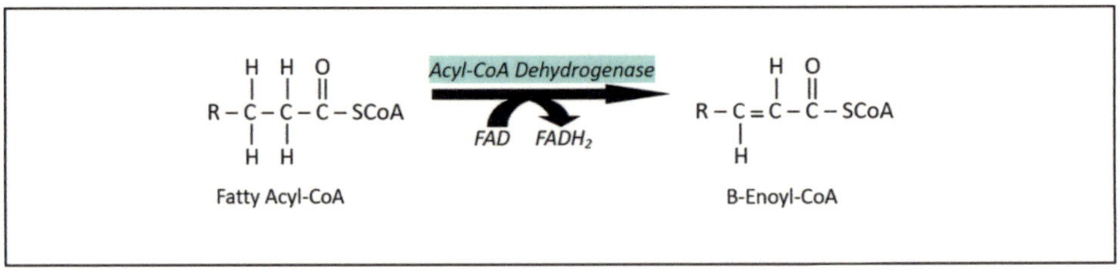

Figure 2.
The dehydrogenation of fatty acyl-CoA to β-enoyl-CoA in the first step of β-oxidation FAD, Flavin Adenine Dinucleotide, FADH2, reduced FAD. The enzynme is highlighted blue.

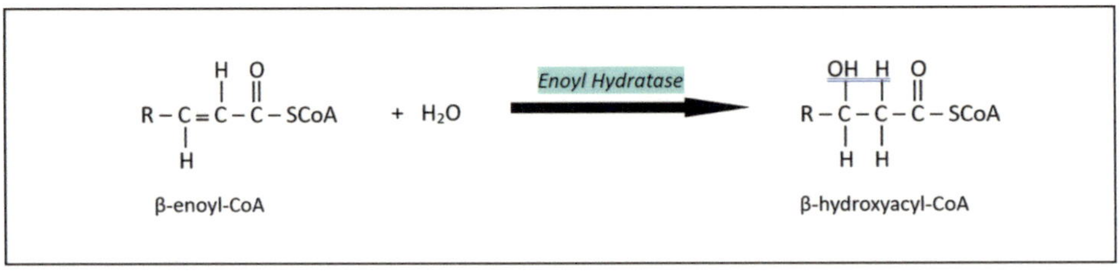

Figure 3.
The hydration of β-enoyl-CoA where a hydroxy-OH group is added to the β carbon thus breaking the double bond and creating β-hydroxyacyl-CoA in the second step of β-oxidation. The enzyme is highlighed blue.

Step 3

The third step is another oxidation reaction catalysed by acyl-CoA dehydrogenase. The 2 hydrogen molecules from the β-carbon hydroxyl group are removed thereby oxidising the β-carbon. This time, nicotinamide adenine dinucleotide (NAD) acts as an electron acceptor. The end product is a β-ketoacyl-CoA, reduced to NADH and H+. Like FADH2, NADH donates the reductive potential to the electron transport chain producing ATP and is recycled in the process (see **Figure 4**) [5, 11, 14].

Step 4

The fourth and final step is a thiolytic reaction catalysed by a β-keto-thiolase enzyme. A CoA molecule is utilised by the lytic enzyme to cleave the bond between the α and β carbon to create acetyl-CoA and a fatty acyl-CoA with 2 less carbon groups (see **Figure 5**) [5, 11, 14].

These four steps are repeated till the fatty acid chain is broken down to acetyl-CoA. The acetyl-CoA molecules produced from β-oxidation can now enter the process of cellular respiration via the citric acid cycle. Odd-chained fatty acids are

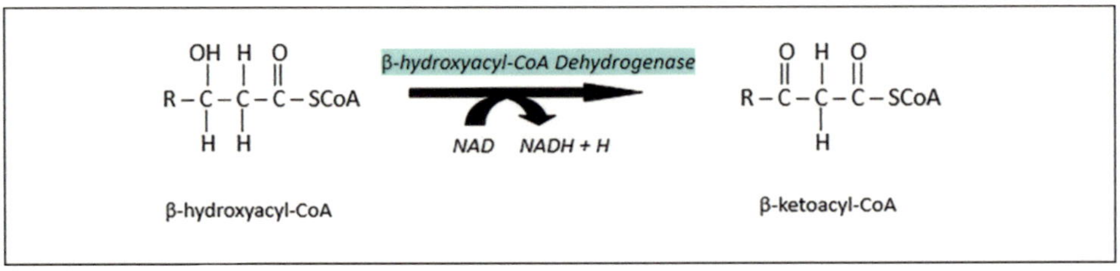

Figure 4.
The oxidation of the β carbon in β-hydroxyacyl-CoA by a dehydrogenase to create β-ketoacyl-CoA in the third step of β-oxidation. The enzyme is highlighted blue.

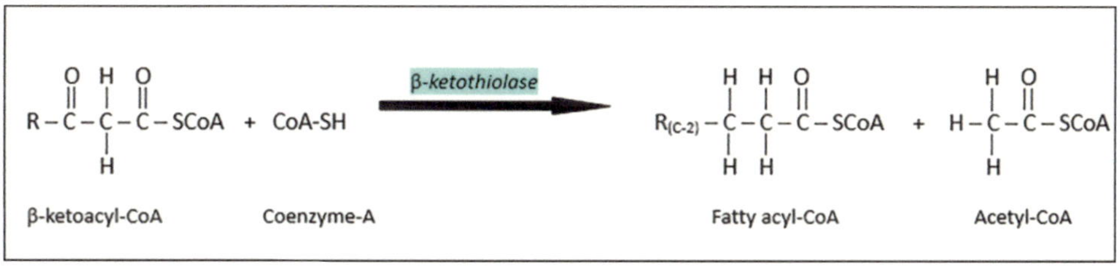

Figure 5.
The final step of β-oxidation is a lytic reation where the bond between the α, β or 2, 3, carbon is cleaved to form acetyl-CoA and re-create fatty acyl-CoA2 carbon groups shorter than that in step 1. The enzyme is highlighted blue.

broken down until it creates one acetyl-CoA and one propionyl-CoA molecule. The propionyl-CoA is then converted enzymatically to succinyl-CoA in a 3-step process. The succinyl-CoA is then used in the citric acid cycle [5].

2.1 The citric acid cycle and electron transport chain

The citric acid cycle takes place in the mitochondrial matrix and is an oxidoreductive process where the bulk of the reduced energy-carrying molecules such as NADH and FADH2 are produced [5, 15]. The citric acid cycle starts with acetyl-CoA being combined with oxaloacetate by citrate synthase with water as a cofactor to form citrate and a free coenzyme-A molecule. The enzyme aconitase then rearranges the molecular structure of citrate to create isocitrate. Next isocitrate is oxidised to α-ketoglutarate by a dehydrogenase enzyme that reduces NAD+ to NADH and also removes a carboxyl group via the byproduct CO2. The α-ketoglutarate is again oxidised by another dehydrogenase enzyme with the cofactor coenzyme-A to reduce another NAD+ to NADH and remove another carboxyl group via the byproduct CO2. This creates succinyl-CoA which acts as a substrate for a phosphorylation reaction via succinyl-CoA synthetase creating Guanine triphosphate (GTP), succinate and coenzyme A. Succinate dehydrogenase, which is in fact complex II on the ETC which we will explore in the next section, then oxidises succinate to fumarate. Fumarate then undergoes a hydration reaction catalysed by fumarase to create malate. In the final step, malate is oxidised by malate dehydrogenase with NAD+ now acting as the electron acceptor and re-forming oxaloacetate which can now bind a new acetyl-CoA molecule and re-enter the citric acid cycle (see **Figure 6**) [5, 15].

So, by way of energy, a single citric acid cycle produces 1 GTP, 1 FADH2, 3 NADH and 2 CO2 molecules. While GTP can be converted to ATP easily, FAHD2 and NADH have to donate their reductive energy to the ETC which generates ATP via oxidative phosphorylation.

2.2 A closer look at oxidative phosphorylation from fatty acid metabolism

There is a lot of ambiguity in literature regarding how the reduced NADH and FAHD2 cofactors transfer electrons to the ETC and carry out oxidative phosphorylation. This can lead to misunderstandings like concluding that FADH2 from β-oxidation in fatty acids donate their protons to the ETC in the same way that FADH2 from the citric acid cycle does, or that FADH2 is the direct substrate which donates the electrons. In reality, the pathways are fundamentally different and FADH2 is a prosthetic group as opposed to the substrate [8, 14, 16]. These

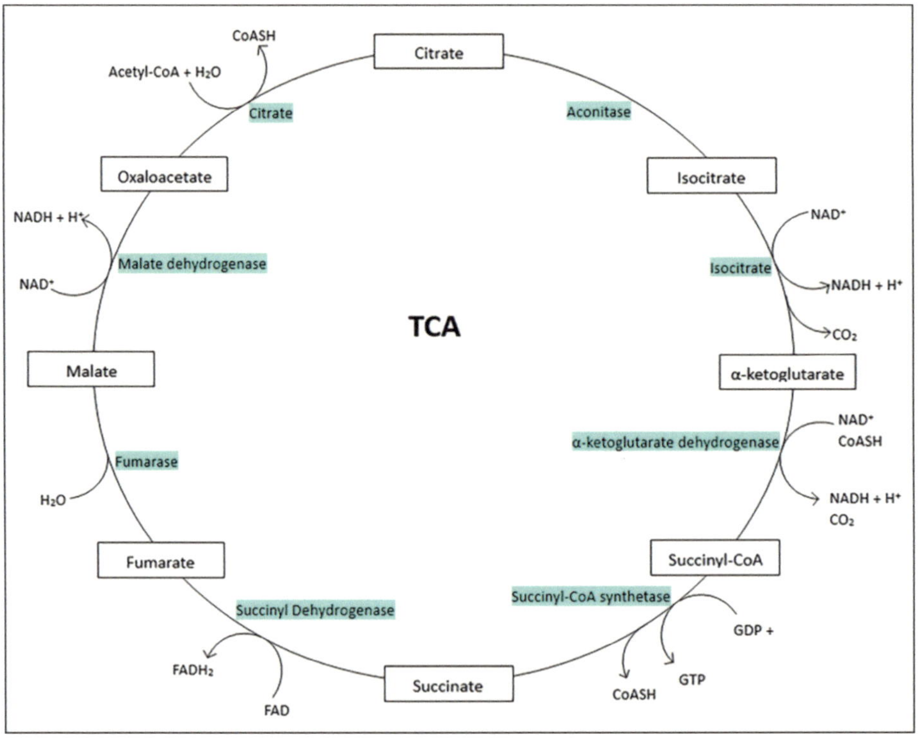

Figure 6.
The citric acid (TCA) cycle. The eight enzymes involved in catalysing each step is highlighted in blue. CoASH, Coenzyme A; NAD⁺, Nicotinamide Adenine Dinucleotide; NADH, reduced NAD⁺,H⁺, hydrogen ion/proiton; CO2, Carbon Dioxice; GDP, Guanine Diphosphate; GTP, Guainine Tripohosphate; FAD, Flavine Adenine Dinucleotide; FADH2, reduced FAD, H2O, water.

differences are vital to understanding how multiple acyl-CoA dehydrogenase deficiency impacts fatty acid metabolism without impacting the ATP generated from the citric acid cycle from acetyl-CoA obtained from other macronutrients like carbohydrates and proteins.

The ETC is located on the inner mitochondrial matrix and consists of a series of proteins along which electrons transfer to create a gradient of protons represented by hydrogen ions (H+). The electron transfer process converges at coenzyme-Q (Q), also known as ubiquinone (UQ), and then follows a linear chain path to its destination. Complex I (CI) of the ETC is the enzyme responsible for oxidising the substrate NADH + H+ created by fatty acid β-oxidation and the citric acid cycle. The electrons in the form of 2(H+ and e-) from the oxidation process by CI are accepted via its non-covalently bound prosthetic group flavin mononucleotide (FMN) reducing it to FMNH2 [16, 17]. Next CI uses the reduced FMH2 to reduce Q/UQ to ubiquinol (UQH2). In C1, the transfer process of this high-energy electron also results in the pumping of H+ from the matrix to the intermembrane space [16, 17]. The process of electron transfer from FADH2 on the other hand varies according to the source of the reduced FADH2 (see **Figure 7**).

The first step of fatty acid β-oxidation is a dehydrogenation reaction carried out by a specific acyl-CoA dehydrogenase (ACAD). To carry out this step, FAD is an essential prosthetic group that attaches non-covalently to ACAD and becomes reduced to FADH2 [8, 14, 16]. Electron transfer flavoproteins (ETF) are unbound mitochondrial

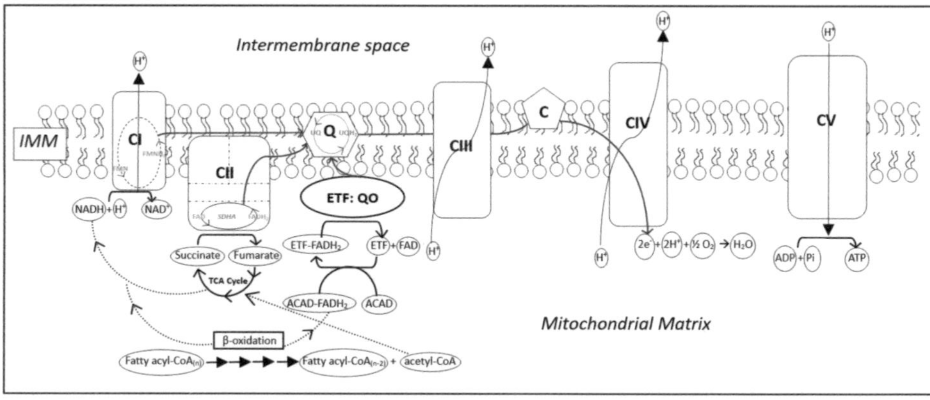

Figure 7.
The culmination of redox reactions from fatty acid metabolism at the electron transport chain. IMM, Inner Mitochondrial Membrane; CI, Complex I; CII, Complex II; CIII, Complex III; CIV, Complex IV; CV, Complex V; Q, coenzyme Q interchangeable with; UQ, ubiquinone; UQH2 ubiquinol; C, cytochrome c; SDHA, succinate dehydrogenase subunit A; SDHB, succinate dehydrogenase subunit B; SDHC, succinate dehydrogenase subunit C; SDHD, succinate dehydrogenase subunit D; TCA cycle, the citric acid cycle; n, number of carbon groups; ETF, Electron Transport Flavoprotein; ETF:QO, Electron Transport Flavoprotein Ubiquinone Oxidoreductase, NAD+, Nicotinamide adenine dinucleotide; NADH, reduced NAD+; H+, hydrogen ion/proton; FAD, Flavine Adenine Dinucleotide; FADH2, reduced FAD; FMN, Flavin Mononucleotide; FMNH2, reduced FMN; e-, electrons; O2, oxygen; H2O, water; ADP, adenosine diphosphate; Pi, inorganic phosphate; ATP, Adenosine Triphosphate.

matrix protein that shuttles the ACAD-associated FADH2 to the membrane bound enzyme electron transfer flavoprotein: ubiquinone oxidoreductase (ETF:QO) [8, 14, 18]. As the name suggests, ETF:QO oxidises FADH2 by removing the 2(H+ and e-) and reduces Q/UQ to UQH2. Thus ETF:QO links fatty acid β-oxidation to the ETC by transferring electrons to Q/UQ bypassing complex II (CII) of the ETC entirely (see **Figure 7**).

As mentioned earlier in citric acid cycle section, succinate dehydrogenase (SHD) is the CII enzyme on the ETC and is made up of 4 different units SDHA, SDHB, SDHC and SDHD [16, 19]. The SDHA subunit is located on the matrix side and is responsible for oxidising succinate in the citric acid cycle to fumarate. This is accomplished by using FAD as the prosthetic group which is bound permanently via a covalent bond to SDHA [16]. The FAD then reduces to FADH2. The SDHC and SDHD subunits embedded in the inner membrane contain the binding site for Q/UQ [16, 19]. The SDHB subunit contains 3 iron–sulphur redox clusters that then catalyse the transfer of 2(H+ and e-) to Q/UQ thereby reducing it to UQH2 [16, 19].

Now that all the pathways from CI, CII and ETF:QO have converged at Q/UQ yielding the reduced UQH2, the pathway for electron transfer follows the same linear chain [14–16]. The electrons and associated protons are then transferred from UQH2 to complex III (CIII) reforming Q/UQ and pumping more H+ from the matrix to the intermembrane space. The electrons are then transferred to cytochrome c (C) with more H+ being pumped by CIII in the process. The last transporter C then delivers the electrons to the final protein complex IV (CIV). The shuttling of electrons from C to CIV allows more H+ to be pumped into the intermembrane space by CIV. The electrons from CIV are then transferred to the matrix side where they combine with oxygen and hydrogen to create water (see **Figure 7**) [14–16].

As a result of the ETC, an electrochemical concentration gradient is created between the intermembrane space and the mitochondrial matrix. When the concentration gradient is sufficient, the H+ ions are pushed through complex V (CV) also known as ATP synthase, causing it to undergo a conformational change that results in adenosine diphosphate (ADP) binding to phosphate group and thus creating ATP [14, 15]. The presence of oxygen is mandatory for the ETC to work, which in turn allows ATP to be synthesised, hence the name oxidative phosphorylation.

3. Multiple acyl-coenzyme a deficiency

Now that we understand the molecular biochemistry of fatty acid metabolism and oxidation, it becomes very clear why multiple acyl Co-A dehydrogenase deficiency (MADD) is a primary fatty acid oxidation disorder. Previously known as glutaric acidaemia type 2, MADD is a rare inborn error of metabolism that occurs most commonly due to mutations in genes encoding ETF subunits, ETF-a and ETF-B, or due to mutations in the electron transfer flavoprotein dehydrogenase (ETFDH) gene which encodes ETF:QO [7, 9, 20]. Ultimately these mutations impair the ability of the FADH2 generated by ACADs in step 1 of β-oxidation to be re-oxidising to FAD. As established, FAD is an essential cofactor for all ACADs and depleting FAD supply has an impact on the oxidation of fatty acids of all lengths causing both upstream and downstream effects [20–22].

The upstream effects stem from the accumulation of fatty acid metabolites. The accumulation of long-chain fatty acids saturates the carnitine shuttle at CPT 1 resulting in a secondary carnitine deficiency and overall elevated levels of serum fatty acylcarnitine of all chain lengths. Fatty acylcarnitine molecules undergo alternative metabolism in peroxisomes and microsomes to create various organic acids that are renally cleared. Overall, the accumulation of fatty acids and it's byproducts results in a high anionic gap metabolic acidosis [7–9, 13, 20].

The downstream effects are due to a lack of acetyl-CoA molecules produced as MADD affects step 1 of the β-oxidation. The lack of acetyl-CoA impairs oxidative phosphorylation via the citric acid cycle where the bulk of the energy shuttle molecules are generated. Acetyl-CoA is also an essential substrate for gluconeogenesis and ketogenesis required in periods of catabolic stress [2–4, 20–21]. The reduced acetyl-CoA and ATP production also reduces ureagenesis resulting in build-up of ammonia [7].

The organs most affected by MADD are those that have a high flux of β-oxidation such as the heart, liver, brain and skeletal muscles [21, 22]. The manifestation of MADD depends on the severity of the mutations, as the resulting ETF and ETF:QO can either be partially functional or redundant. There are 3 subtypes of MADD, sutype 1 and 2 are neonatal variations both with and without congenital abnormalities respectively. Both Subtypes 1 and 2 are detected on the Newborn Screening test and are fatal within the first few weeks of life due to rapidly progressing cardiomyopathies [21, 23]. Subtype 3 of MADD is also known as late-onset MADD and is the least fatal but also the hardest to detect. Individuals with late-onset MADD have mutations that result in proteins with some residual enzymatic activity with symptoms only manifesting during periods of catabolic stress [20–23]. Even then the manifestation depends on the severity of the catabolic stress, how exhausted their glycogen stores are and how reliant the body is on fatty acid metabolism for energy. There is a range from vague symptoms like hypotonia, muscle weakness and exercise intolerance to more severe features of hepatomegaly, encephalopathy and cardiomyopathy that can lead to comas and death [7, 21–23].

When suspected, the expected biochemical abnormalities on tests would be high anionic gap metabolic acidosis, hypoketotic hypoglycaemia, hyperammonemia, elevated plasma acylcarnitine profile and a urine metabolic screen positive for organic acids [7, 20–23]. If diagnosed properly, it is easily treatable with avoidance of prolonged fasting and a modified low-protein, low-fat diet with carnitine and riboflavin supplementation. The role of riboflavin supplementation is to increase mitochondrial supply of FAD. It is important to note that there are more mutations being discovered in other genes such as FLAD1 that have a role in MADD with the list of possible causative gene mutations likely being far from exhausted [9, 24]. Understanding the molecular differences in how fatty acids and downstream metabolites are metabolised is crucial to shedding light on the pathophysiology of MADD and various other fatty acid oxidation disorders and hence crucial to informing appropriate treatment.

Conflict of interest

The authors declare no conflict of interest.

Author details

Naini Nishita Rao
Canberra Health Service, Canberra, Australia

*Address all correspondence to: n.nishita.rao@gmail.com

References

[1] Murray B, Rosenbloom C. Fundamentals of glycogen metabolism for coaches and athletes. Nutrition Reviews. 2018;**76**(4):243-259. DOI: 10.1093/nutrit/nuy001

[2] Alghannam AF, Ghaith MM, Alhussain MH. Regulation of energy substrate metabolism in endurance exercise. International Journal of Environmental Research and Public Health. 2021;**18**(9):4963. DOI: 10.3390/ijerph18094963

[3] Browning JD, Baxter J, Satapati S, Burgess SC. The effect of short-term fasting on liver and skeletal muscle lipid, glucose, and energy metabolism in healthy women and men. Journal of Lipid Research. 2012;**53**(3):577-586. DOI: 10.1194/jlr.P020867

[4] Jensen NJ, Wodschow HZ, Nilsson M, Rungby J. Effects of ketone Bodies on Brain metabolism and function in neurodegenerative diseases. International Journal of Molecular Sciences. 2020;**21**(22):8767. DOI: 10.3390/ijms21228767

[5] Houten SM, Violante S, Ventura FV, Wanders RJ. The Biochemistry and physiology of mitochondrial fatty acid β-oxidation and its genetic disorders. Annual Review of Physiology. 2016;**78**:23-44. DOI: 10.1146/annurev-physiol-021115-105045

[6] Grice AS, Peck TE. Multiple acyl-CoA dehydrogenase deficiency: A rare cause of acidosis with an increased anion gap. British Journal of Anaesthesia. 2001;**86**(3):437-441. DOI: 10.1093/bja/86.3.437

[7] Wanders RJ, Waterham HR, Ferdinandusse S. Metabolic interplay between peroxisomes and other subcellular organelles including mitochondria and the endoplasmic reticulum. Frontiers in Cell and Developmental Biology. 2016;**3**:83. DOI: 10.3389/fcell.2015.00083

[8] Schönfeld P, Wojtczak L. Short- and medium-chain fatty acids in energy metabolism: The cellular perspective. Journal of Lipid Research. 2016;**57**(6):943-954. DOI: 10.1194/jlr.R067629

[9] Wanders RJA, Visser G, Ferdinandusse S, Vaz FM, Houtkooper RH. Mitochondrial fatty acid oxidation disorders: Laboratory diagnosis, pathogenesis, and the complicated route to treatment. Journal of Lipid and Atherosclerosis. 2020;**9**(3):313-333. DOI: 10.12997/jla.2020.9.3.313

[10] Casale J, Huecker MR. Fasting [Internet]. 2020. Available from: https://www.ncbi.nlm.nih.gov/books/NBK534877/ [Accessed: 03 February 2024]

[11] Chandel NS. Lipid Metabolism. Cold Spring Harbor perspectives in Biology. 2021;**13**(9):a040576. DOI: 10.1101/cshperspect.a040576

[12] Baltscheffsky M, Nyrén P. The synthesis and utilization of inorganic pyrophosphate. In: Ernster L, editor. New Comprehensive Biochemistry. Amsterdam: Elsevier; 1984. pp. 187-206. DOI: 10.1016/S0167-7306(08)60316-6

[13] Longo N, Frigeni M, Pasquali M. Carnitine transport and fatty acid oxidation. Biochimica et Biophysica Acta. 2016;**1863**(10):2422-2435. DOI: 10.1016/j.bbamcr.2016.01.023

[14] Henriques BJ, Katrine Jentoft Olsen R, Gomes CM, Bross P. Electron transfer flavoprotein and its role in mitochondrial energy metabolism in health and disease. Gene. 2021;**776**:145407. DOI: 10.1016/j.gene.2021.145407

[15] Martínez-Reyes I, Chandel NS. Mitochondrial TCA cycle metabolites control physiology and disease. Nature Communications. 2020;**11**(102):1-11. DOI: 10.1038/s41467-019-13668-3

[16] Gnaiger E. Complex II ambiguities —— FADH2 in the electron transfer system. Journal of Biological Chemistry. 2023;**300**(1):105470. DOI: 10.1016/j.jbc.2023.105470

[17] Curtabbi A, Enríquez JA. The ins and outs of the flavin mononucleotide cofactor of respiratory complex I. IUBMB Life. 2022;**74**(7):629-644. DOI: 10.1002/iub.2600

[18] Watmough NJ, Frerman FE. The electron transfer flavoprotein: Ubiquinone oxidoreductases. Biochimica et Biophysica Acta. 2010;**1797**(12):1910-1916. DOI: 10.1016/j.bbabio.2010.10.007

[19] Bandara AB, Drake JC, Brown DA. Complex II subunit SDHD is critical for cell growth and metabolism, which can be partially restored with a synthetic ubiquinone analog. BMC Molecular and Cell Biology. 2021;**22**(1):35. DOI: 10.1186/s12860-021-00370-w

[20] Przyrembel H, Wendel U, Becker K, et al. Glutaric aciduria type II: Report on a previously undescribed metabolic disorder. Clinica Chimica Acta. 1976;**66**(2):227-239. DOI: 10.1016/0009-8981(76)90060-7

[21] Grünert SC. Clinical and genetical heterogeneity of late-onset multiple acyl-coenzyme a dehydrogenase deficiency. Orphanet Journal of Rare Diseases. 2014;**9**(1):117. DOI: 10.1186/s13023-014-0117-5

[22] Vockley J. Specific Fatty Acid Oxidation Disorders [Internet]. 2023. Available from: https://www.uptodate.com/contents/specific-fatty-acid-oxidation-disorders#:~:text=Fatty%20acid%20oxidation%20disorders%20(FAODs,of%20the%20fatty%20acid%20chain. [Accessed: 01 February 2024]

[23] Prasun P. Multiple acyl-CoA dehydrogenase deficiency. In: Adam MP, Feldman J, Mirzaa GM, Pagon RA, Wallace SE, Bean LJH, Gripp KW, Amemiya A, editors. GeneReviews. Seattle: National Library Of Medicine; 2020. Available from: https://www.ncbi.nlm.nih.gov/books/NBK558236/

[24] Yubero D, Brandi N, Ormazabal A, Garcia-Cazorla A, Perez-Duenas B, Campistol J, et al. Targeted next generation sequencing in patients with inborn errors of metabolism. PLoS One. 2016;**11**(5):e0156359. DOI: 10.1371/journal.pone.0156359

Chapter 4

Oxidative Stress-Induced Ferroptosis

Shuang Shang and Lifang Ma

Abstract

Oxidative stress is viewed as a cause of damage to proteins, DNAs, and lipids, therefore inducing alteration in their function and ultimately leading to cellular damage. Lipid peroxidation often occurs under oxidative damage conditions. A high rate of lipid peroxidation can cause cell death, such as apoptosis, necrosis, and ferroptosis. Different from apoptosis and autophagy, ferroptosis is a kind of regulated cell death (RCD) that features the dysfunction of lipid peroxide, resulting in tumors, inflammatory, and cardiovascular diseases. Extensive studies suggest that ferroptosis plays a pivotal role in some human diseases, thus providing novel opportunities for therapy. We focus on the physiological and pathological mechanisms of oxidative stress and ferroptosis and finally discuss the prospect and challenge of therapeutic strategies toward ferroptosis in several diseases.

Keywords: oxidative stress, ferroptosis, antioxidants, lipid peroxidation, cancer, cardiovascular diseases

1. Introduction

Oxidative stress was described as a kind of imbalance between oxidant and antioxidant systems. It is characterized by the weakened ability of the endogenous system to defend against the oxidative attack toward several target biomolecules [1]. The field of oxidative stress research includes biochemistry, physiology, pathophysiology, and cell biology, furthermore, medicine, health, and disease research may also involve oxidative stress research. Oxide metabolism complements the details of hydroperoxide metabolism in mammalian organisms and the relationship in bioenergetics [2].

Both reactive oxygen and nitrogen species can induce oxidative stress, which is traditionally viewed as a cause of damage to proteins, DNAs, and lipids, therefore inducing alteration in their function and eventually leading to cellular damage [3]. Proteins are mainly functional biomolecules that drive cellular activity. Oxidative damage to proteins may lead to protein dysfunction [4]. DNA damage can be caused by reactive oxygen species (ROS) radical and may do harm to mutations, epigenetic changes, and genetic instability [5, 6]. Oxidative stress can damage several types of oncogenes and antioncogenes and then cause mutations that are known to trigger cancer. Lipid peroxidation often occurs under the condition of oxidative damage and it usually locates on the cellular membrane, finally results in disability of membrane properties [7]. Many other molecules may also be damaged by the reactive end products of lipid peroxidation. Low lipid peroxidation leads to the adaption of cell defense mechanisms, while a higher rate of lipid peroxidation can cause cell death, such as apoptosis, necrosis, and proptosis.

The field of ferroptosis has seen great growth in the past few years. Ferroptosis is a kind of regulated cell death (RCD) that was found to have a close connection to iron and lipid metabolism [8]. It features the clearing dysfunction of lipid peroxide, the existence of redox-active iron, and the oxidation of polyunsaturated fatty acid (PUFA)-containing phospholipids [9]. This modality of cell death has been proved to be regulated by multiple cellular metabolic pathways, including iron handling, redox homeostasis, mitochondrial activity, and metabolism of many other molecules such as amino acids and lipids. Furthermore, it also exists in various signaling pathways linked to disease. Although the idiographic function of ferroptosis remains poorly researched, its connection with diseases, such as tumors, inflammatory, and cardiovascular diseases (CVDs), has been widely discussed.

From invertebrates to vertebrates, oxidative stress seems to play an important role in functioning as a regulator in both physiological and pathological mechanisms. Here, we will summarize the features, production process, function, and mechanisms of oxidative stress, discuss its relevance with lipid peroxidation and ferroptosis, and raise some unsolved issues of ferroptosis.

2. Generation of reactive oxygen species and antioxidant system

2.1 Generation process of ROS

ROS are radicals, ions, or molecules and are some kinds of the most familiar oxidants in cells. ROS can be divided into two categories: free oxygen radical ROS and non-radical ROS, which are formed by the partial reduction of molecular oxygen to superoxide, hydrogen peroxide, lipid peroxides, or the corresponding hydroxyl and peroxyl radicals.

There are also endogenous and exogenous ROS. Endogenous ROS are often produced by mitochondria, peroxisomes, and inflammatory cells [10]. Mitochondria are important cellular organelles and major sites of the production of ROS. When mitochondrial reactive oxygen species (mtROS) accumulate, p38 has been proven to be activated, then leading to paraquat (PQ)-induced proptosis [11]. The production of mtROS has also been identified to be essential for signaling in antigen-specific T cell activation, while dysfunction of mitochondria in T cells has been seen as feature of some autoimmune diseases in humans [12]. ROS production in peroxisomes can be regulated by post-translational modifications [13]. In recent years, diverse evidence has indicated that peroxisomes are important cellular sources of various signaling molecules, including ROS. Peroxisomes take part in different processes of high physiological importance and might be crucial in cellular redox homeostasis maintenance. Inflammation is a basic defensive response toward harmful stimulation, but the overactivation of inflammatory responses is related to most kinds of human diseases [14]. Inflammatory cells can produce ROS through the reduction of the molecular oxygen. ROS at a proper level can function as critical signaling molecules in the regulation of various physiological functions, including inflammatory responses. However, overproduced ROS have a toxic impact on cells and can straightforwardly oxidize biological molecules, such as proteins, DNAs, and lipids, further exacerbating the development of inflammatory responses and causing various inflammatory diseases.

ROS have been reported to be closely linked with inflammation and tumor survival and invasion, furthermore, leading to the happening of tumor development, cardiovascular disorders, neurodegenerative diseases, and other pathologies.

2.2 Antioxidant systems

Antioxidant systems have been greatly researched in recent years for their protective roles in oxidative stress caused by ROS. They have been defined as substances that have the ability to reduce oxidant species, which means lower oxidative stress, decreased DNA mutations, and less cell damage [15].

There are several classification criteria, such as enzymatic and non-enzymatic, endogenous and exogenous. The earliest type of antioxidant system identified as a mechanism against oxidative stress is featured by preventing the occurrence of ROS and blocking and capturing radicals that have formed. Another important type of antioxidant system in cells is identified as a repairing process, which can eliminate the damaged biomolecules before alteration of cell metabolism [16]. However, with continued free radical action, the capacity of the defensive systems against ROS can be reduced, inducing the occurrence of some human diseases [17].

When cells are exposed to harmful ROS, a series of reactions and induces activation of internal defensive mechanisms (enzymatic or non-enzymatic) will be caused. Several antioxidants, which can remove reactive species and derivatives, have been explored for their beneficial effects against oxidative stress. Antioxidant systems consist of enzymatic and non-enzymatic antioxidants, but the main burden of antioxidant defense is shouldered by enzymatic antioxidants.

2.2.1 Enzymatic antioxidants

The major role in antioxidant defense is conducted by antioxidant enzymes, not by the small molecular antioxidant compounds. Superoxide dismutase (SOD), glutathione reductase (GR), and catalases (CAT) are well-known antioxidant enzymes, which provide effective protection against oxidative stress [18].

SODs were first reported half a century ago by McCord and Fridovich. It is a metalloenzyme known as one of the most effective enzymatic antioxidants found in subcellular compartments and protects cells from the toxic effects of overproduced ROS [19]. SODs can decompose superoxide into oxygen and hydrogen peroxide and help prevent the transition metal-catalyzed formation of hydroxyl radicals. According to different metal cofactors, there are three isoforms of SOD enzymes, categorized as Cu/Zn-SODs, Mn-SODs, and Fe-SODs [20]. There also are extracellular SODs which can prevent endothelial cells from dysfunction by protecting against NO inactivation.

Glutathione (GSH)/glutaredoxin (GRX) system is one of the main cellular antioxidant pathways. In the GSH/GRX system, glutathione reductase functions to maintain the supply of reduced glutathione that acts importantly in the balance of cellular reactive oxygen species and then can further reduce oxidized GRX [21, 22]. In addition, it can transform toxic substances such as hydroxy peroxide into nontoxic metabolites in body homeostasis. GR is known as one of the most abundant reducing thiols in most cells and is known for its excellent antioxidant properties. Therefore, it can have a significant impact on cancer development.

Catalase has been studied extensively. It can be divided into several categories: bifunctional catalases, monofunctional catalases, and manganese-containing catalases [23]. CAT is an enzyme that has a tetrameric heme group and is an antioxidant enzyme that can convert hydrogen peroxide to water and oxygen rapidly. It has been proven to be effective on some toxic compounds via peroxidative reaction. CAT has been treated as a cancer therapeutic which has been widely studied to play a role in reducing oxidative stress and hypoxia in tumor microenvironment (TME) [24], both of which are hypothesized to decrease tumor growth.

2.2.2 Non-enzymatic antioxidants

The non-enzymatic antioxidants are characterized by molecules that can rapidly inactivate oxidants and radicals, including GSH and uric acid (UA).

GSH is composed of three amino acids: cysteine, glycine, and glutamic acid. In physiological conditions, it exists in many different tissues and acts as an antioxidant [25]. GSH in the human body is present in several redox forms, among which the most famous are reduced GSH and oxidized glutathione (GSSG). GSH in different cells has cellular type-special concentration and role. Besides being an effective antioxidant, it has many other functions not related to defense against ROS, for example, it also works in the repair processes of cellular damage.

Uric acid, chemically characterized as an antioxidant in the human plasma, both correlates and predicts the development of hypertension, obesity, and CVDs, conditions associated with oxidative stress [26]. It is also widely accepted that intracellular uric acid can defense oxidative stress via preventing ROS generated by xanthine oxidase. In prior studies, uric acid suppressed ROS accumulation and protected against ischemic neuronal injury [27]. Furthermore, it can also protect the erythrocyte membrane from lipid oxidation by eliminating free radicals.

3. Lipid peroxidation and ferroptosis

3.1 Lipid peroxidation

An increasing number of research have implicated lipid peroxides as key mediators in some pathological conditions, including inflammation, neurodegenerative disease, cancer, as well as cardiovascular diseases. In addition, lipid peroxidation is one of the crucial downstream features of ferroptosis, which is identified as a novel form of non-apoptotic RCD. Lipid peroxidation can be classified into two categories: lipid endoperoxides and lipid hydroperoxides. Lipid endoperoxides have been extensively researched as the pivotal intermediates in the generation of prostaglandins while recently, lipid hydroperoxides have also been considered as the key mediators in cell death and diseases [28].

3.2 Relationships between ferroptosis and other types of cell death

Ferroptosis is a new type of oxidative stress-dependent RCD, associated with iron overload and lipid peroxidation, then resulting in the accumulation of lipid hydroperoxides which can cause cell death.

3.2.1 Ferroptosis and autophagy

Autophagy is an evolutionarily conserved cellular process that is capable of degrading various long-lived proteins and biological molecules through the lysosomal pathway [29]. Autophagy has been extensively studied as a pivotal cellular response, and it is described to be involved in many diseases. Recently, ferroptosis was found to depend on autophagy. An increasing number of studies focused on the relationship between ferroptosis and autophagy. Recently, it has been suggested that ferroptosis and autophagy can synergize with each other.

The overactivation of autophagy has been confirmed to be induced by the ferroptosis inducers erastin [30]. Furthermore, autophagy triggers iron-dependent ferroptosis by degradation and release of ferritin and induction of the expression of transferrin receptor 1(TFR1) [31–33]. Remarkably, nuclear receptor coactivator 4(NCOA4)-mediated autophagy promotes ferritin phagocytosis to degrade and then release ferric ions, leading to the accumulation of autophagic vesicles. Therefore, NCOA4 overexpression can increase ferritin degradation via transporting excess ferric ions into the cytoplasm, causing oxidative stress responses and inducing ferroptosis in cells. Deficiency of autophagy causes decreased intracellular iron and reduced lipid peroxidation, which can prevent cell death from erastin-induced ferroptosis. As a result, induction of autophagy-dependent ferroptosis has also been suggested as a possible antineoplastic strategy.

These findings provide a novel understanding of the interplay between ferroptosis and autophagy. However, this interactive relationship still remains controversial, and the autophagy-inducing mediator is unknown.

3.2.2 Ferroptosis and apoptosis

Apoptosis is a kind of RCD and is considered a vital component of various physiological processes, such as normal cell turnover and development of immune systems [34]. In normal cells, apoptosis occurs when there is serious damage or protein misfolding, including an imbalance in apoptotic factors and genetic mutations. While in tumor cells, the genetic variations result in the molecular misfunction that inhibits cell division and even induces apoptosis, thus unlimited cell division and cell death evasion exist.

P53 is an important tumor suppressor and nuclear transcription factor that regulates the expression of genes. It can regulate cell biological processes, promote ROS production, and trigger ferroptosis and apoptosis [35]. ROS may act as a bridge between apoptosis and ferroptosis [36]. There has been a study suggesting that the C/EBP-homologous protein/p53 upregulated modulator of apoptosis (CHOP/PUMA) axis can respond to ferroptosis inducers [37], which means the axis may play a role in ferroptosis agent-mediated sensitization to apoptosis induced by tumor necrosis factor-related apoptosis-inducing ligand (TRAIL). Erastin and artesunate (ART) induced endoplasmic reticulum (ER) stress and also promoted the expression of PUMA via CHOP. Furthermore, cysteine dioxygenase type 1(Cdo1) can regulate cysteine metabolism, thereby affecting the capacity of cellular antioxidative, which makes Cdo1 in an important situation in promoting both ferroptosis and apoptosis [38]. It promotes ferroptosis via decreasing the number of antioxidants, leading to the cell membrane auto-peroxidation by the Fenton reaction. Cdo1 also can facilitate apoptosis mainly through the production of cysteine metabolism, taurine, and decreased levels of antioxidants. It has been proved that the altered function of Cdo1 can be exhibited in many cancers, emphasizing its crucial role in tumor cell survival.

Evidence of the direct regulation of ferroptosis by p53 is still lacking. Thus, the specific regulatory mechanism still needs to be further studied.

3.3 Biological pathways involved in ferroptosis

It is widely accepted that excessive accumulation of iron can generate ROS through the Fenton reaction, resulting in lipid peroxidation and ferroptosis. The mechanism of ferroptosis is still unclear, but there are some hypotheses in recent research.

Through the initial study of the role of system Xc--GSH-GPX4 pathway in suppressing ferroptosis, phospholipid hydroperoxides (PLOOHs) have been identified as executioners of ferroptosis [39]. Overaccumulation of PLOOHs may possibly cause rapid and unrepairable damage to cell membranes, then leading to cell death. Glutathione peroxidase 4(GPX4) is one of the most well-known regulatory factors in

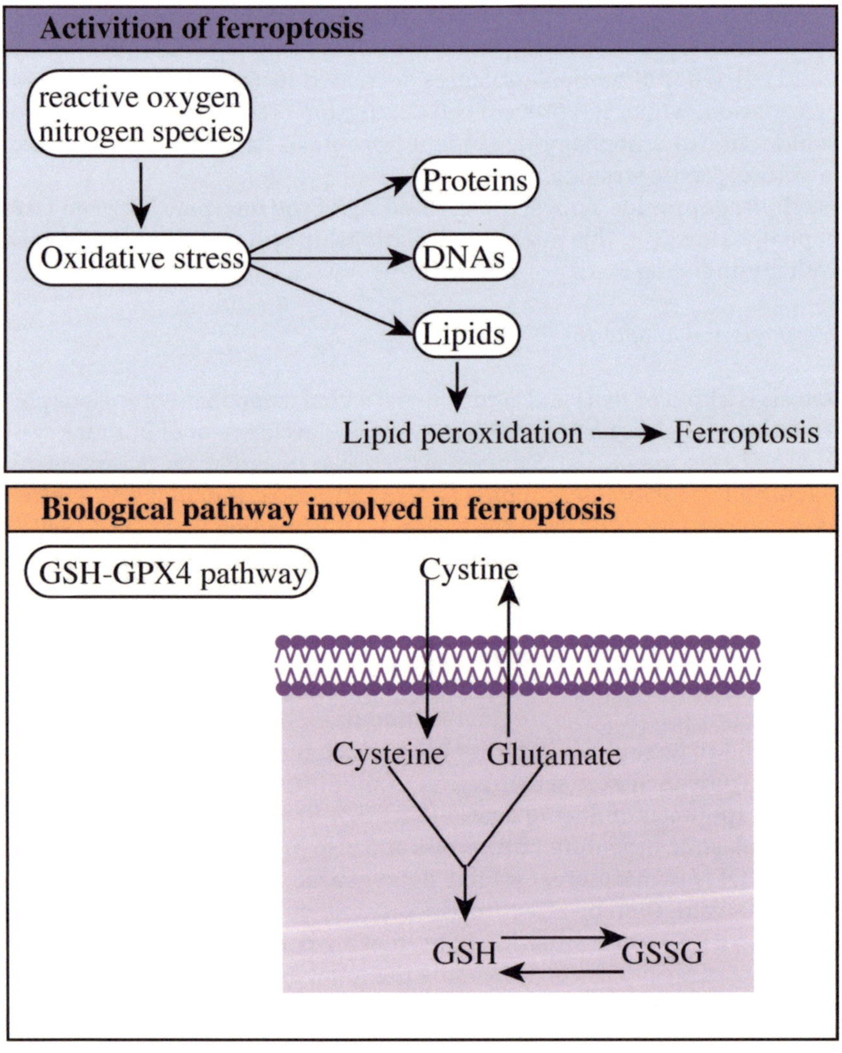

Figure 1.
The mechanism of ferroptosis.

the ferroptosis signaling pathways protecting cells from ferroptosis. GPX4 can inhibit the Fenton reaction by catalyzing the oxidation of GSH and removing lipid peroxides [40]. It is also a major PLOOH-neutralizing enzyme. Direct or indirect inactivation of GPX4 is a classic induction mechanism in ferroptosis. Previous studies revealed a general mechanism underlying erastin- or RSL3-induced ferroptosis. Both compounds can inactivate GPX4-RSL3. RSL3 can directly inactivate or deplete GPX4 to induce ferroptosis. Erastin inhibits the system Xc- and indirectly inactivates GPX4, resulting in the accumulation of lipid peroxides that promote ferroptosis [41].

In addition to the classical GPX4-dependent ferroptosis pathway, there are two newly discovered protecting pathways independent of GPX4, including the ferroptosis suppressor protein (FSP1)-ubiquinone (CoQ10) pathway [42] and guanosine triphosphate cyclohydrolase 1(GCH1) pathway [43]. With the help of NADPH, FSP1 can inhibit lipid peroxidation and ferroptosis by converting CoQ10 into ubiquinol, which can reduce lipid radicals directly [44, 45]. GCH1 can regulate the synthesis of tetrahydrobiopterin (BH4). Ferroptosis can be inhibited by selectively preventing the oxidation of lipids via GCH1-mediated BH4 production (**Figure 1**) [46].

4. The potential therapeutic implications in diseases

Oxidative stress is associated with a great number of pathologies and is identified as one of the most primary causes of pathology. Basically, the contribution of oxidative stress to the etiology of those pathologies can be grouped into the following types: one is that oxidative stress is the primary cause of pathology; the other is it is the secondary contributor to different disease progressions, such as in inflammation, cancer, and CVDs. However, the actual role of oxidative stress in diseases remained unclear, this categorization above is tentative.

It has been proved that in many human diseases, oxidative stress often occurs secondary to the pathology caused by other factors. Oxidative stress can interfere with various signaling pathways and have effects on multiple biological processes via promoting inflammation, modifying proteins, inducing cell death, and many other biological mechanisms. These effects usually can accelerate pathological progression and worsen the symptoms of diseases.

In physiological conditions, the balance between oxidant and antioxidant compounds moderately improves prooxidants to produce mild oxidative stress, which means that the endogenous antioxidant systems are needed to intervene in the oxidative damage [47]. Strategies to enhance these defenses are principal to fundamental antioxidant therapy. With age growing, when the imbalance of endogenous antioxidants and repair systems happens, the issue of oxidative stress will become more acute. Therefore, various measures preventing or inhibiting these aggressive factors are aimed at decreasing disease incidence. Most studies toward the induction of antioxidant enzymes mainly focused on the regulatory mechanisms, the significance in diseases, and potential therapeutic inducers.

4.1 Cancer

ROS is involved in almost all phases of tumorigenesis, including transformation and growth of tumor cells, proliferation, invasion, and metastasis [48]. On the other hand, oxidative stress can also induce apoptosis and ferroptosis, and reduce the chance for transformation of tumor cells and thus prevent tumorigenesis. Therefore, oxidative stress is considered the main mechanism of radiation and chemotherapeutic drugs [49] and it is involved in almost all stages of cancer. Cancer cells can produce more ROS than normal cells. Therefore, cancer cells are exposed to increased oxidative stress in the loci. It can be observed that the oxidative markers increase in various cancers. For example, patients with non-small cell lung cancer (NSCLC) may exhale more H_2O_2 than normal individuals [50, 51]. Furthermore, compared to matched normal tissues, in prostate cancers [52] and lung cancers [53, 54], 8-hydroxy-2′-deoxyguanosine2 (8-OHdG) was observed to be significantly elevated, while it was also detected to elevate in breast cancer tissues [55].

In recent years, there has been emerging evidence indicating that ferroptosis-inducing agents toward ferroptosis-induced cell death in cancer may provide a promising strategy for cancer therapy [56]. Ferroptosis has been proved by Extensive studies that it plays a role in tumor suppression, thus providing novel opportunities for cancer therapy. Sorafenib can be described as a multiple kinase inhibitor, and it was reported to have the ability to induce ferroptosis and exert anti-tumor effects in many kinds of tumor cell lines, such as NSCLC [57], hepatocellular carcinoma cells [58], and human kidney cancer cells [59]. While we try to describe oxidative stress as harmful to the human body, it is true that it is exploited as a therapeutic approach to treat clinical conditions such as cancer, with a certain degree of clinical success.

4.2 Inflammation

Basically, many researchers have noticed that ROS and antioxidants can influence the immune systems in the human body. Inflammation is tightly connected with oxidative stress, which can trigger a series of transcription factors such as nuclear factor-kappa B1 (NF-κB1) [60], nuclear factor E2-related factor 2(Nrf2) [61], and pro-inflammatory cytokines tumor necrosis factor-alpha (TNF-α) [62], thus leading to the different expressions of chemokines, inflammatory cytokines, and anti-inflammatory molecules.

The correlation between chronic inflammation and oxidative stress has already been confirmed. For example, in asthma [63] and allergic rhinitis [64], the imbalance between the activity of oxidative species, the promotors of oxidative damages, and the antioxidative defense is involved. Oxidative stress disrupts the cell signaling pathways and impairs arachidonic acid (AA) metabolism, thus enhancing airway and systemic inflammation [65]. It was also stated that allergic conditions can be initiated by the generation of TH1/TH2 cytokines [66], which are associated with oxidative stress-induced inflammation. Pterostilbene (Pts) alleviates oxidative stress and allergic airway inflammation through the regulation of AMPK/Sirt1and Nrf2/HO-1 signaling pathways [67]. Lycopene has high antioxidative activity, and there are results indicating that a daily dose of lycopene can exert an in vivo antioxidative effect against exercise-induced asthma (EIA) in some patients [68].

Redundant ROS induces oxidative stress and then consumes antioxidants in cells, which may further aggravate the production of lipid peroxidation and inflammatory responses [69]. Lipid peroxidation, in turn, can drive the increase of modified low-density lipoprotein (LDL), which can promote inflammation to some extent via macrophage polarization. Therefore, lipid peroxidation can be observed in many kinds of physiological conditions in different diseases, such as cell death and inflammatory responses in the pathophysiology. Aldose reductase (AR) can catalyze the limiting step of polyol pathway of glucose metabolism [70, 71]. Except for the ability to reduce glucose to sorbitol, it also can reduce lipid peroxidation by deriving aldehydes and glutathione conjugates. Previous studies suggested that catalytic activity of AR plays a role in various inflammatory diseases such as sepsis, asthma, and atherosclerosis.

4.3 Cardiovascular diseases

Many studies have mentioned that iron was involved in the development of atherosclerosis (AS) by affecting the level of lipid peroxidation in vivo. Ferroptosis connected with lipid peroxidation and iron deposition is prominent in the progression of AS [72]. Ferritin and LDL-cholesterol levels showed a synergistic relationship with the incidence of CVDs [73, 74]. Iron-catalyzed free radical reactions generate ox-LDL in endothelial cells, smooth muscle cells, and macrophages, which are important risk factors in atherosclerotic lesions [75]. Transferrin is the major iron-binding molecule in plasma which has a mutual effect on transferrin receptor protein 1(TFR1). It may deliver extracellular Fe^{3+} into cells, leading to iron overload and increased cell susceptibility to ferroptosis [76]. Therefore, ferroptosis in epithelial cells, vascular smooth muscle cells, and macrophages may be deduced to be related to the destabilization of atherosclerotic plaque. As a result, lowering iron levels in plasma can be an effective intervention to prevent iron overload and atherosclerosis progression.

Stroke induces the interruption or reduction of brain blood in blood circulation, which is one major cause of death in the world. Ferroptosis has been proven to be involved in pathological cell damage in stroke [77]. As reported before, iron plays a part in many physiological processes, such as oxidative reaction, erythropoiesis, and immunity. It also consists of cytochromes a-c in the oxidative chain and adenosine triphosphate (ATP). However, disturbances of iron homeostasis lead to neuronal damage after ischemic injury [78, 79]. Iron leads to ferroptosis and neuronal injury by converting superoxide and hydrogen peroxides into reactive hydroxyl radicals which may do harm to cells and tissues [80]. As a result, iron is causally associated with lipid hydroperoxide production and disease incidence of ischemic stroke. Both basic and clinical research results reveal that dysregulation of lipid peroxidation induces stroke [81]. There is also a clinical study exhibiting that the average levels of lipid hydroperoxides in the plasma of stroke patients were higher than controls [82].

CVDs usually begin with vascular disorders and end with heart failure (HF), which is the common final clinical result of CVDs. It was demonstrated that ferroptosis is highly correlated with terminally differentiated cardiomyocyte death. Some studies showed that the increased levels of free iron pool and lipid peroxide indicate that ferroptosis is directly related to HF [83]. For example, a study has shown that ferroptosis plays a vital role in doxorubicin (DOX)-induced HF [84, 85]. DOX results in the degradation of cardiac heme and release of free iron via the accumulation of oxidized phospholipids in cells and overexpression of heme oxygenase 1(HO-1), which causes ferroptosis in cardiomyocytes, finally leading to HF. Therefore, inhibitors of ferroptosis or HO-1 can significantly prevent HF caused by DOX. Another study has shown the decreased levels of ferritin heavy chain 1(FTH1) in HF and that cardiomyocyte death can be caused by iron deposition and an increase in oxidative stress [86]. Decreased levels of FTH1 in cells will lead to disordered iron metabolism and accumulation of ROS, thus causing ferroptosis. So, regulating the upstream axis and thus altering the level of FTH1 can be a key regulatory strategy in treating HF.

These studies revealed that ferroptosis is related to the regulation of HF. Therefore, it is a promising therapy to trap ferroptosis to treat ferroptosis-associated cardio disorders.

4.4 Neurodegenerative disorders

Oxidative stress has been mentioned in the occurrence and progress of several neurodegenerative disorders, such as Alzheimer's disease (AD) and Parkinson's disease (PD) [87–89]. Oxidative stress was widely accepted to modify the inflammatory response, including neuroinflammation. It can be inferred that oxidative stress and neuroinflammation are linked and can affect each other.

AD is defined by a gradual deterioration of cognitive capacities which can result from synapses degradation and neuron death, especially in the hippocampus [90]. ROS is generated when there is a disproportion between antioxidants and oxidants and then the processes mentioned above will primarily originate and even enhance. The reason for this can either be a free radical enhancement or a reduction in the defense of antioxidants. OS induced by ROS is also considered a critical factor in the pathogenesis of AD because of its relationship with the accumulation and deposition of β-amyloid [91]. PD happens when the substantia nigra pars compacta of the brain exhibit dopaminergic neuron loss [90]. ROS is generated mainly in mitochondria in the neurons and neuroglia cells. The main reasons why ROS is overproduced in the PD are neuroinflammation, mitochondrial dysfunction, increased levels of iron and calcium, etc. The degradation of dopaminergic neurons is connected with the overproduction and extreme accumulation of ROS, which may be related to both mitochondrial dysfunctions and inflammation [91]. Although the utter process that determines dopaminergic neuronal loss is not clearly researched, it has been

Disease	Relationship with ferroptosis	Potential therapeutic target
Lung cancer	RBMS1 regulates the translation of SLC7A11, reduces SLC7A11-mediated cystine uptake, and then promotes ferroptosis.	Targeting ferroptosis pathway is a potential strategy for the treatment of lung cancer.
Acute Myocardial Infarction	Oxidized phosphatidylcholine-containing phospholipids(OxPL) are generated during AMI and have negative effects on cardiomyocyte viability. Lipid peroxidation was found to be a key factor leading to oxidative damage of cardiomyocytes.	The use of therapeutics Targeting ferroptosis and using cyclosporine A together can be a promising strategy toward AMI.
Heart Failure	During heart failure, FTH expression is downregulated, and a large amount of ferrous ions are released, ultimately leading to ROS accumulation and iron death.	Overexpression of SLC7A11 in cardiomyocytes increases glutathione levels and prevents FTH-induced ferroptosis.
Atherosclerosis	The accumulation of lipid peroxides, restriction of glutathione synthesis, and disturbance of iron homeostasis are related to ferroptosis and then induce atherosclerosis.	Activating NRF2 and inhibiting ROS release and iron levels can inhibit ferroptosis.
Alzheimer's disease	NOX4 induces ferroptosis in nerve cells by oxidative stress-induced lipid peroxidation via the damage of mitochondrial metabolism. Intracellular accumulation of amyloid beta(Aβ) can also induce ferroptosis in nerve cells.	Targeting iron and ferroptosis could be a promising therapeutic option for AD.
Parkinson's disease	Aggregate-membrane interaction can induce ferroptosis. α-synuclein oligomers further induce lipid peroxidation.	GSH and its related molecules may have neuroprotective effects in PD pathology.

Table 1.
Diseases caused by ferroptosis and potential therapeutic targets.

suggested that ROS can be a key factor. However, there is still no exact therapy for PD. But to understanding the mechanisms of ROS related to PD's evolution can aid the development of treatments (**Table 1**).

5. Conclusions and perspectives

Lipid peroxidation is one of the crucial downstream features of ferroptosis, leading to the happen of tumor development, CVDs, and other pathologies. We

Number	Recent study about ROS and ferroptosis
1	Snap25 attenuates neuronal injury by reducing ferroptosis in acute ischemic stroke. Si, W., Sun, B., Luo, J., Li, Z., Dou, Y., & Wang, Q. Experimental neurology, 367, 114,476. DOI: 10.1016/j.expneurol.2023.114476
2	Cancer-associated fibroblasts impair the cytotoxic function of NK cells in gastric cancer by inducing ferroptosis via iron regulation. Yao, L., Hou, J., Wu, X., Lu, Y., Jin, Z., Yu, Z., Yu, B., Li, J., Yang, Z., Li, C., Yan, M., Zhu, Z., Liu, B., Yan, C., & Su, L. Redox biology, 67, 102,923. DOI: 10.1016/j.redox.2023.102923
3	Quercetin Protects against MPP+/MPTP-Induced Dopaminergic Neuron Death in Parkinson's Disease by Inhibiting Ferroptosis. Lin, Z. H., Liu, Y., Xue, N. J., Zheng, R., Yan, Y. Q., Wang, Z. X., Li, Y. L., Ying, C. Z., Song, Z., Tian, J., Pu, J. L., & Zhang, B. R. Oxidative medicine and cellular longevity, 2022, 7,769,355. DOI: 10.1155/2022/7769355
4	Phospholipids with two polyunsaturated fatty acyl tails promote ferroptosis. Qiu, B., Zandkarimi, F., Bezjian, C. T., Reznik, E., Soni, R. K., Gu, W., Jiang, X., & Stockwell, B. R. Cell, 187(5), 1177–1190.e18. DOI: 10.1016/j.cell.2024.01.030
5	APE1 inhibition enhances ferroptotic cell death and contributes to hepatocellular carcinoma therapy. Du, Y., Zhou, Y., Yan, X., Pan, F., He, L., Guo, Z., & Hu, Z. Cell death and differentiation, 31(4), 431–446. DOI: 10.1038/s41418-024-01270-0
6	A guideline on the molecular ecosystem regulating ferroptosis. Dai, E., Chen, X., Linkermann, A., Jiang, X., Kang, R., Kagan, V. E., Bayir, H., Yang, W. S., Garcia-Saez, A. J., Ioannou, M. S., Janowitz, T., Ran, Q., Gu, W., Gan, B., Krysko, D. V., Zhu, X., Wang, J., Krautwald, S., Toyokuni, S., Xie, Y., ... Tang, D. Nature cell biology, 10.1038/s41556-024-01360-8. Advance online publication. DOI: 10.1038/s41556-024-01360-8
7	GAS41 modulates ferroptosis by anchoring NRF2 on chromatin. Wang, Z., Yang, X., Chen, D., Liu, Y., Li, Z., Duan, S., Zhang, Z., Jiang, X., Stockwell, B. R., & Gu, W. Nature communications, 15(1), 2531. DOI:10.1038/s41467-024-46,857-w
8	Identification of a targeted ACSL4 inhibitor to treat ferroptosis-related diseases. Huang, Q., Ru, Y., Luo, Y., Luo, X., Liu, D., Ma, Y., Zhou, X., Linghu, M., Xu, W., Gao, F., & Huang, Y. Science advances, 10(13), eadk1200. DOI:10.1126/sciadv.adk1200

Table 2.
Recent study about ROS and ferroptosis.

summarize the features, production process, function, and mechanisms of oxidative stress and relevant biological processes in this review. An in-depth understanding of the molecular mechanisms underlying lipid peroxidation-induced ferroptosis is provided. It is widely accepted that excessive accumulation of iron can generate ROS through the Fenton reaction, resulting in lipid peroxidation and ferroptosis. The mechanism of ferroptosis is still unclear, but there are some hypotheses in previous research which can help us to understand signaling pathways and defense mechanisms of ferroptosis. It has been proved that oxidative stress can interfere with various signaling pathways and have effects on multiple biological processes. The usage of antioxidants in prevention and treatment is still in dispute. While we try to describe oxidative stress as harmful to the human body, it is true that it is exploited as a therapeutic approach to treat clinical conditions such as cancer, with a certain degree of clinical success. The link between ferroptosis and therapeutic strategies is emerging, but the exact mechanism remains to be further observed experimentally. Further studies should explore how ferroptosis is tightly regulated to provide therapeutic strategies for targeting ferroptosis to treat human diseases, offering guidance in developing new targeted antioxidant drugs (**Table 2**).

Acknowledgements

This study was supported by the National Natural Science Foundation of China (82273139), Shanghai Science and Technology Committee Rising-Star Program (22QA1408300), Excellent Talents Nurture Project of Shanghai Chest Hospital (2021YNZYY02).

Author details

Shuang Shang and Lifang Ma*
Department of Clinical Laboratory, Shanghai Chest Hospital, Shanghai Jiao Tong University School of Medicine, Shanghai, China

*Address all correspondence to: mlf2281@sjtu.edu.cn

References

[1] Pisoschi AM, Pop A. The role of antioxidants in the chemistry of oxidative stress: A review. European Journal of Medicinal Chemistry. 2015;**97**:55-74

[2] Chance B, Sies H, Boveris A. Hydroperoxide metabolism in mammalian organs. Physiological Reviews. 1979;**59**(3):527-605

[3] Li H et al. Role of Nrf2 in the antioxidation and oxidative stress induced developmental toxicity of honokiol in zebrafish. Toxicology and Applied Pharmacology. 2019;**373**:48-61

[4] Hawkins CL, Davies MJ. Detection, identification, and quantification of oxidative protein modifications. The Journal of Biological Chemistry. 2019;**294**(51):19683-19708

[5] Shi T et al. DNA damage and oxidant stress activate p53 through differential upstream signaling pathways. Free Radical Biology & Medicine. 2021;**172**:298-311

[6] Valko M et al. Free radicals, metals and antioxidants in oxidative stress-induced cancer. Chemico-Biological Interactions. 2006;**160**(1):1-40

[7] Girotti AW, Korytowski W. Trafficking of oxidative stress-generated lipid hydroperoxides: Pathophysiological implications. Free Radical Research. 2023;**57**(2):130-139

[8] Angeli JPF et al. Ferroptosis inhibition: Mechanisms and opportunities. Trends in Pharmacological Sciences. 2017;**38**(5):489-498

[9] Friedmann Angeli JP, Krysko DV, Conrad M. Ferroptosis at the crossroads of cancer-acquired drug resistance and immune evasion. Nature Reviews. Cancer. 2019;**19**(7):405-414

[10] Sullivan LB, Chandel NS. Mitochondrial reactive oxygen species and cancer. Cancer & Metabolism. 2014;**2**:17

[11] Chen K et al. Mitochondrial reactive oxygen species initiate gasdermin D-mediated pyroptosis and contribute to paraquat-induced nephrotoxicity. Chemico-Biological Interactions. 2024;**390**:110873

[12] Chen J et al. Mitochondrial reactive oxygen species and type 1 diabetes. Antioxidants & Redox Signaling. 2018;**29**(14):1361-1372

[13] Del Río LA, López-Huertas E. ROS generation in peroxisomes and its role in cell Signaling. Plant & Cell Physiology. 2016;**57**(7):1364-1376

[14] Liu J et al. Reactive oxygen species (ROS) scavenging biomaterials for anti-inflammatory diseases: From mechanism to therapy. Journal of Hematology & Oncology. 2023;**16**(1):116

[15] Gulcin İ. Antioxidants and antioxidant methods: An updated overview. Archives of Toxicology. 2020;**94**(3):651-715

[16] Cheeseman KH, Slater TF. An introduction to free radical biochemistry. British Medical Bulletin. 1993;**49**(3):481-493

[17] Godic A et al. The role of antioxidants in skin cancer prevention and treatment. Oxidative Medicine and Cellular Longevity. 2014;**2014**:860479

[18] Veen L et al. Dietary antioxidants, non-enzymatic antioxidant capacity and the risk of osteoarthritis in the Swedish National March Cohort. European Journal of Nutrition. 2021;**60**(1):169-178

[19] Gill SS, Tuteja N. Reactive oxygen species and antioxidant machinery in abiotic stress tolerance in crop plants. Plant Physiology and Biochemistry. 2010;**48**(12):909-930

[20] Wen F, Tan ZG, Xiang J. Cu-Zn SOD suppresses epilepsy in pilocarpine-treated rats and alters SCN2A/Nrf2/HO-1 expression. Epileptic Disorders. 2022;**24**(4):647-656

[21] Couto N, Wood J, Barber J. The role of glutathione reductase and related enzymes on cellular redox homoeostasis network. Free Radical Biology & Medicine. 2016;**95**:27-42

[22] Brzozowa-Zasada M et al. Glutathione reductase expression and its prognostic significance in colon cancer. International Journal of Molecular Sciences. 2024;**25**(2):1097

[23] Chelikani P, Fita I, Loewen PC. Diversity of structures and properties among catalases. Cellular and Molecular Life Sciences. 2004;**61**(2):192-208

[24] Sheen A et al. Tumor-localized catalases can fail to alter tumor growth and transcriptional profiles in subcutaneous syngeneic mouse tumor models. Redox Biology. 2023;**64**:102766

[25] Forman HJ, Zhang H, Rinna A. Glutathione: Overview of its protective roles, measurement, and biosynthesis. Molecular Aspects of Medicine. 2009;**30**(1-2):1-12

[26] Kimura Y, Tsukui D, Kono H. Uric acid in inflammation and the pathogenesis of atherosclerosis. International Journal of Molecular Sciences. 2021;**22**(22):12394

[27] Sautin YY, Johnson RJ. Uric acid: The oxidant-antioxidant paradox. Nucleosides, Nucleotides & Nucleic Acids. 2008;**27**(6):608-619

[28] Gaschler MM, Stockwell BR. Lipid peroxidation in cell death. Biochemical and Biophysical Research Communications. 2017;**482**(3):419-425

[29] Lv X et al. The crosslinks between ferroptosis and autophagy in asthma. Frontiers in Immunology. 2023;**14**:1140791

[30] Park E, Chung SW. ROS-mediated autophagy increases intracellular iron levels and ferroptosis by ferritin and transferrin receptor regulation. Cell Death & Disease. 2019;**10**(11):822

[31] Zhou B et al. Ferroptosis is a type of autophagy-dependent cell death. Seminars in Cancer Biology. 2020;**66**:89-100

[32] Hou W et al. Autophagy promotes ferroptosis by degradation of ferritin. Autophagy. 2016;**12**(8):1425-1428

[33] Lee S et al. Autophagy mediates an amplification loop during ferroptosis. Cell Death & Disease. 2023;**14**(7):464

[34] Elmore S. Apoptosis: A review of programmed cell death. Toxicologic Pathology. 2007;**35**(4):495-516

[35] Mao C et al. A G3BP1-interacting lncRNA promotes Ferroptosis and apoptosis in cancer via nuclear sequestration of p53. Cancer Research. 2018;**78**(13):3484-3496

[36] Su LJ et al. Reactive oxygen species-induced lipid peroxidation in apoptosis, autophagy, and Ferroptosis. Oxidative

Medicine and Cellular Longevity. 2019;**2019**:5080843

[37] Hong SH et al. Molecular crosstalk between ferroptosis and apoptosis: Emerging role of ER stress-induced p53-independent PUMA expression. Oncotarget. 2017;**8**(70):115164-115178

[38] Chen X, Poetsch A. The role of Cdo1 in Ferroptosis and apoptosis in cancer. Biomedicine. 2024;**12**(4):918

[39] Jiang X, Stockwell BR, Conrad M. Ferroptosis: Mechanisms, biology and role in disease. Nature Reviews. Molecular Cell Biology. 2021;**22**(4):266-282

[40] Xie Y et al. GPX4 in cell death, autophagy, and disease. Autophagy. 2023;**19**(10):2621-2638

[41] Dixon SJ et al. Ferroptosis: An iron-dependent form of nonapoptotic cell death. Cell. 2012;**149**(5):1060-1072

[42] Chen H et al. Ferroptosis and its multifaceted role in cancer: Mechanisms and therapeutic approach. Antioxidants (Basel). 2022;**11**(8):1504

[43] Jiang Y et al. CircLRFN5 inhibits the progression of glioblastoma via PRRX2/GCH1 mediated ferroptosis. Journal of Experimental & Clinical Cancer Research. 2022;**41**(1):307

[44] Bersuker K et al. The CoQ oxidoreductase FSP1 acts parallel to GPX4 to inhibit ferroptosis. Nature. 2019;**575**(7784):688-692

[45] Koppula P et al. A targetable CoQ-FSP1 axis drives ferroptosis- and radiation-resistance in KEAP1 inactive lung cancers. Nature Communications. 2022;**13**(1):2206

[46] Hu Q et al. Blockade of GCH1/BH4 Axis activates Ferritinophagy to mitigate the resistance of colorectal cancer to Erastin-induced Ferroptosis. Frontiers in Cell and Development Biology. 2022;**10**:810327

[47] Yoshikawa T, You F. Oxidative stress and bio-regulation. International Journal of Molecular Sciences. 2024;**25**(6):3360

[48] Hayes JD, Dinkova-Kostova AT, Tew KD. Oxidative stress in cancer. Cancer Cell. 2020;**38**(2):167-197

[49] Conklin KA. Chemotherapy-associated oxidative stress: Impact on chemotherapeutic effectiveness. Integrative Cancer Therapies. 2004;**3**(4):294-300

[50] Galiniak S, Rohovyk N, Rachel M. Biomarkers of nitrosative stress in exhaled breath condensate and serum among patients with cystic fibrosis. Advances in Medical Sciences. 2023;**68**(2):202-207

[51] Chan HP et al. Elevated levels of oxidative stress markers in exhaled breath condensate. Journal of Thoracic Oncology. 2009;**4**(2):172-178

[52] Di Minno A et al. 8-Hydroxy-2-Deoxyguanosine and 8-Iso-prostaglandin F2α: Putative biomarkers to assess oxidative stress damage following robot-assisted radical prostatectomy (RARP). Journal of Clinical Medicine. 2022;**11**(20):6102

[53] Orfanakos K et al. The predictive value of 8-Hydroxy-Deoxyguanosine (8-OHdG) serum concentrations in irradiated non-small cell lung carcinoma (NSCLC) patients. Biomedicine. 2024;**12**(1):134

[54] Shen J et al. 8-Hydroxy-2′-deoxyguanosine (8-OH-dG) as a potential survival biomarker in patients with nonsmall-cell lung cancer. Cancer. 2007;**109**(3):574-580

[55] Nour Eldin EEM et al. 8-Hydroxy-2′-deoxyguanosine as a discriminatory biomarker for early detection of breast cancer. Clinical Breast Cancer. 2019;**19**(2):e385-e393

[56] Zhang C et al. Ferroptosis in cancer therapy: A novel approach to reversing drug resistance. Molecular Cancer. 2022;**21**(1):47

[57] Groenendijk FH et al. Sorafenib synergizes with metformin in NSCLC through AMPK pathway activation. International Journal of Cancer. 2015;**136**(6):1434-1444

[58] Tang W et al. The mechanisms of sorafenib resistance in hepatocellular carcinoma: Theoretical basis and therapeutic aspects. Signal Transduction and Targeted Therapy. 2020;**5**(1):87

[59] Lee J, Park SH. Anti-cancer activity of microbubble conjugated with Sorafenib containing liposome and IL4R-targeting peptide in kidney cancer cells. Cellular and Molecular Biology (Noisy-le-Grand, France). 2023;**69**(14):266-271

[60] Alharbi KS et al. Nuclear factor-kappa B and its role in inflammatory lung disease. Chemico-Biological Interactions. 2021;**345**:109568

[61] Sajadimajd S, Khazaei M. Oxidative stress and cancer: The role of Nrf2. Current Cancer Drug Targets. 2018;**18**(6):538-557

[62] Li J et al. TNF-α inhibitors with anti-oxidative stress activity from natural products. Current Topics in Medicinal Chemistry. 2012;**12**(13):1408-1421

[63] Michaeloudes C et al. Molecular mechanisms of oxidative stress in asthma. Molecular Aspects of Medicine. 2022;**85**:101026

[64] Han M et al. Oxidative stress and antioxidant pathway in allergic rhinitis. Antioxidants (Basel). 2021;**10**(8):1266

[65] Luo Y et al. Role of arachidonic acid lipoxygenase pathway in asthma. Prostaglandins & Other Lipid Mediators. 2022;**158**:106609

[66] Ke X et al. Quercetin improves the imbalance of Th1/Th2 cells and Treg/Th17 cells to attenuate allergic rhinitis. Autoimmunity. 2023;**56**(1):2189133

[67] Xu C et al. Pterostilbene suppresses oxidative stress and allergic airway inflammation through AMPK/Sirt1 and Nrf2/HO-1 pathways. Immunity, Inflammation and Disease. 2021;**9**(4):1406-1417

[68] Neuman I, Nahum H, Ben-Amotz A. Reduction of exercise-induced asthma oxidative stress by lycopene, a natural antioxidant. Allergy. 2000;**55**(12):1184-1189

[69] Hörl G et al. In vitro oxidation of LDL by ozone. Chemistry and Physics of Lipids. 2014;**183**:18-21

[70] Thakur S et al. Aldose reductase: A cause and a potential target for the treatment of diabetic complications. Archives of Pharmacal Research. 2021;**44**(7):655-667

[71] Ramana KV, Srivastava SK. Aldose reductase: A novel therapeutic target for inflammatory pathologies. The International Journal of Biochemistry & Cell Biology. 2010;**42**(1):17-20

[72] Li N et al. Ferroptosis and its emerging roles in cardiovascular diseases. Pharmacological Research. 2021;**166**:105466

[73] El-Gebali HH et al. Lipid peroxidative damage in the erythrocytes

and elevation of serum LDL-cholesterol, apolipoprotein-B, ferritin and uric acid with age and in coronary heart disease patients. Saudi Medical Journal. 2000;**21**(2):184-189

[74] de Valk B, Marx JJ. Iron, atherosclerosis, and ischemic heart disease. Archives of Internal Medicine. 1999;**159**(14):1542-1548

[75] Guo Q, Qian C, Qian ZM. Iron metabolism and atherosclerosis. Trends in Endocrinology and Metabolism. 2023;**34**(7):404-413

[76] Feng H et al. Transferrin receptor is a specific Ferroptosis marker. Cell Reports. 2020;**30**(10):3411-3423.e7

[77] Xu Y et al. Role of Ferroptosis in stroke. Cellular and Molecular Neurobiology. 2023;**43**(1):205-222

[78] Liang D, Minikes AM, Jiang X. Ferroptosis at the intersection of lipid metabolism and cellular signaling. Molecular Cell. 2022;**82**(12):2215-2227

[79] Li C et al. Ferroptosis contributes to hypoxic-ischemic brain injury in neonatal rats: Role of the SIRT1/Nrf2/GPx4 signaling pathway. CNS Neuroscience & Therapeutics. 2022;**28**(12):2268-2280

[80] Guo J, Tuo QZ, Lei P. Iron, ferroptosis, and ischemic stroke. Journal of Neurochemistry. 2023;**165**(4):487-520

[81] Lee WC et al. Lipid peroxidation dysregulation in ischemic stroke: Plasma 4-HNE as a potential biomarker? Biochemical and Biophysical Research Communications. 2012;**425**(4):842-847

[82] Delgado-Martín S, Martínez-Ruiz A. The role of ferroptosis as a regulator of oxidative stress in the pathogenesis of ischemic stroke. FEBS Letters. Published online April 26, 2024

[83] Savarese G et al. Iron deficiency and cardiovascular disease. European Heart Journal. 2023;**44**(1):14-27

[84] Xiong J, Zhou R, Deng X. PRDX6 alleviated heart failure by inhibiting doxorubicin-induced ferroptosis through the JAK2/STAT1 pathway inactivation. In Vitro Cellular & Developmental Biology-Animal. 2024;**60**:354-364

[85] Ma Y et al. Echinacoside ameliorates doxorubicin-induced cardiac injury by regulating GPX4 inhibition-induced ferroptosis. Experimental and Therapeutic Medicine. 2024;**27**(1):29

[86] Chen X et al. Role of TLR4/NADPH oxidase 4 pathway in promoting cell death through autophagy and ferroptosis during heart failure. Biochemical and Biophysical Research Communications. 2019;**516**(1):37-43

[87] Song S et al. ALOX5-mediated ferroptosis acts as a distinct cell death pathway upon oxidative stress in Huntington's disease. Genes & Development. 2023;**37**(5-6):204-217

[88] Chen Z, Zhong C. Oxidative stress in Alzheimer's disease. Neuroscience Bulletin. 2014;**30**(2):271-281

[89] Trist BG, Hare DJ, Double KL. Oxidative stress in the aging substantia nigra and the etiology of Parkinson's disease. Aging Cell. 2019;**18**(6):e13031

[90] Liu Z et al. Oxidative stress in neurodegenerative diseases: From molecular mechanisms to clinical applications. Oxidative Medicine and Cellular Longevity. 2017;**2017**:2525967

[91] Lin MT, Beal MF. Mitochondrial dysfunction and oxidative stress in neurodegenerative diseases. Nature. 2006;**443**(7113):787-795

Chapter 5

Oxidative Stress and Male Subfertility: Recent Detection Methodologies

Petros Karkalousos, Maria Trapali and George Albert Karikas

Abstract

The high concentrations of ROS/RNS/RSS free radicals and neutral compounds have a negative effect on human fertility, both in men and women. The overall result is known as oxidative stress, which also impacts male infertility and has been confirmed in both animal models and infertile men by basic semen analysis. Determination of oxidative stress is not a routine test, but its consequences are diagnosed indirectly by the low values of basic semen parameters and the poor sperm function and by assessing the man's overall lifestyle. According to a plethora of epidemiology/clinical data, oxidative stress could be reduced by radical lifestyle interventions such as antioxidant nutrition intake, weight loss, and smoking and alcohol cessation. This chapter presents the reactive species generation; their crucial relation/mechanisms with body disorders/diseases, in general; and more specific lab measurements on human sperm (e.g., decrease of basic semen analysis, increase of sperm DNA fragmentation and sperm apoptosis). Special mention will be made toward the trace elements Zn, Se, and Cu's importance in male reproductive system.

Keywords: fertility, sperm, oxidative stress, reactive species, DNA fragmentation, apoptosis, trace elements

1. Introduction

1.1 Free radicals and oxidative stress

Oxygen is often referred to as "*Janus gas,*" the two-faced ancient Greco-Roman god, as it exhibits both positive benefits and potentially harmful side effects on biological systems. Within the body, oxygen is involved, among others, in the structure of various molecules, known as "reactive oxygen species" (ROS). ROS are oxygen-negative ions (O_2^{-2}, O_2^-) or more complex ions with other chemical elements (OH^-, RO^-, ROO^-). All above components include also electrically neutral organic compounds (i.e., ROOH, HOCl). ROS are a key component of oxidative stress and are related to a group of free radicals and neutral molecules that oxidize macromolecules in the organism [1]. Oxidative radical species also include nitrogen molecules (NO, NO_2, HNO_4) and nitrogen derivatives (ozone), which are called reactive nitrogen species (RNS). Recently, sulfur ions (RSR^-, $GSSG^-$) as well as sulfur neutral compounds (SO_2, SO_3) have also been found to belong in the above group, called reactive sulfur species (RSS). Inside an organism, free radicals are converted, under appropriate conditions, into the corresponding neutral compounds. Thus, the term reactive species (RS) includes all ROS, RNS, and RSS, while the term free radicals (FR) includes all the negative ions of ROS, RNS, and RSS (**Table 1**).

Chemical formula	Name of free radicals	Formation, metabolic reactions
Part 1: Free radicals		
O_2^{-2}	Peroxide anion	A single oxygen-oxygen covalent bond. It is produced by oxidation (−1) of oxygen atoms.
O_2^-	Superoxide anion	One-electron reduction state of O_2, formed in many antioxidation reactions and by the electron transport chain in mitochondria. Undergoes dismutation to form H_2O_2 spontaneously or by enzymatic catalysis and is a precursor for metal-catalyzed OH formation.
HOO^-	Hydroperoxide radical	Produced by H_2O_2.
OH^-	Hydroxyl radical	Three-electron reduction state, formed by Fenton reaction and decomposition of peroxynitrite ($ONOO^-$). Extremely reactive, attacking most cellular components.
RO^- & ROO^-	Alkoxide and peroxy radical	Oxygen-centered organic radicals. Lipid forms participate in lipid peroxidation reactions. Produced in the presence of oxygen by radical addition to double bonds or hydrogen abstraction.
RSR^-	Radical Sulfur species	Produced by reaction between ROS and thiol.
$GSSG^-$	Glutathionyl radical	Produced during redox signaling reactions.
$ONOO^-$	Peroxinitrite radical	Formed in a rapid reaction between O_2^- and NO. Lipid soluble and similar in reactivity to hypochlorous acid.
Part 2: Non-free radicals		
NO_2 & NO	Nitrogen dioxide & nitrogen monoxide	NO converts to NO_2 and HNO_4 (See HNO_4). NO_2 is an irritating, toxic gas, not biodegradable and precursor of ozone.
H_2O_2	Hydrogen peroxide	Two-electron reduction state, formed by dismutation of O_2^- or by direct reduction of O_2. Lipid soluble and thus able to diffuse between membranes.
ROOH	Organic hydroperoxides	Formed by radical reaction with cellular components such as lipids and nucleobases.
$^1O_2$	Singlet oxygen	Singlet oxygen is a non-radical involved in cholesterol oxidation, which can be accelerated by the co-presence of fatty acid methyl ester.
O_3	Ozone	Mainly formed by chemical reactions between volatile organic compounds (VOCs) and oxides of nitrogen (NOx), in the presence of sunlight and higher temperatures. The human-caused sources of VOCs and nitrogen oxides are industrial. O_3 is harmful in high concentration.

Chemical formula	Name of free radicals	Formation, metabolic reactions
HOCl	Hypochlorous acid	Formed from H_2O_2 by myeloperoxidase. Lipid soluble and highly reactive. It readily oxidizes protein constituents, including thiol groups, amino groups, and methionine.
HNO_4	Peroxynitric acid	Formed by $ONOO^-$. It is formed by protonation of $ONOO^-$. It can undergo homolytic cleavage to form hydroxyl radical (OH-) and nitrogen dioxide (NO_2).
SO_2, SO_3	Sulfur dioxide & Sulfur trioxide	Are produced by H_2SO_4. SO_3 is unstable.

Table 1.
The most important elements of oxidative stress (ROS, RSS, RNS) [1–3].

Depending on their concentration, the RS act either beneficially or harmfully within the organism. RS are produced in all body cells and take part in key pathways of cellular metabolism. In general, FR (**Table 1**, part 1) act via their charge bind to counter ions of various macromolecules of the organism, thus substantially influencing their function. Their conversion to the electrically neutral compounds (**Table 1**, part 2) inhibits their function, but they can be converted back to free radicals if specific chemical conditions promote the conversion. The importance of RSS in the overall oxidative stress activity was first reported in 2001 and documented relatively recently. RSS are formed by the sulfur effect on ROS and depending on conditions can exhibit oxidative or reductive activity on proteins [4].

The general public knows RS mainly for their pathological activity, since they attack transmembrane cellular proteins, lipids, and DNA molecules. The harmful oxidative "attack" of ROS/RSS/RNS molecules on the organism macromolecules is characterized as "oxidative distress." In contrast, the beneficial effect of the active compounds is called "oxidative eustress." Their high concentration and chronic action leads to oxidative stress, which plays a central and promoting role in the pathophysiology of many different disorders/diseases, including complications of pregnancy, as well as, a decrease in male fertility (**Table 2**) [7]. For instance, mitochondrial ROS are involved in high-energy electron transfers and support the production of large amounts of adenosine-5-triphosphate (ATP) via oxidative phosphorylation. In addition to the production of ATP molecules, which provide the necessary energy to the body, ROS have a beneficial role in the body's defense against microorganisms and in cell signaling [8]. ROS/RSS/RNS produced by metabolic processes and the human environment (diet, pollutants, etc.) are also an important source of ROS (**Table 3**).

1.2 RS action in the body

FR were discovered in 1900 [10], while their oxidizing activity was discovered many years later [11]. FR's strong activity is considered incompatible with life, but in the 1950s, it was shown that FRs are also present in biological systems [12]. In 1985, Sies formulated the first definition of oxidative stress [13]. Numerous scientific papers followed concerning its impact on various biological systems, including male and female reproductive organs.

As already stated, the normal concentration of RS has a beneficial role in cell signaling, apoptosis, gene expression, and ion transport in the cells of the organism. On the contrary, high RS have a negative binding effect on very crucial macromolecules. The most important action of RS is redox signaling through post-translational

Organ	Disease
Brain	Alzheimer
	Parkinson
	Memory loss
	Depression
Lung	Asthma
	Chronic bronchitis
Cardiovascular	Atherosclerosis
	Hypertension
	Cardiomyopathy
	Ischemia
Articulations	Rheumatism
	Arthritis
Kidney	Renal disease
Gastrointestinal tract	Inflammatory bowel disease
	Peptic ulcers
Multi-organs	Cancer
	Inflammation
	Diabetes
	Aging
	Infection
Genitalia	Decrease of spermatogenesis (oligoasthenoteratospermia)
	Decrease in the ability of sperm to fertilize eggs
	Decrease of follicular fluid quality

Table 2.
Main diseases promoted by oxidative stress [5, 6].

Internal sources/pathways & molecules	**External sources**
Cellular metabolism/Mitochondria	Exercise
Xanthine oxidase	Cigarette smoke
Peroxisomes	Environmental pollutants (drugs, pesticides, transition metals)
Phagocytosis	Food (lipids, carbohydrates, highly processed food)
Arachidonate pathways	Radiation
Drug metabolites	Industrial solvents
Internal sources/diseases	Ozone
Ischemia/reperfusion injury	Pathogens (bacteria, virus, fungus, parasite)
Anxiety	

Table 3.
Basic production sources of RS in humans [9].

modifications. The main redox signaling agents are hydrogen peroxide (H_2O_2) and superoxide anion radical ($O_2^{-\cdot}$), which are under the control of growth factors, cytokines, and enzymes (mainly NADPH oxidases and mitochondrial electron transport chain enzymes) [14]. Among all RS, the most active is H_2O_2, acting in concentrations of *nmole*. Nitric oxide, hydrogen sulfide, and so on are also involved in redox signaling. Recent methodological developments have allowed the evaluation of molecular interactions of specific RS molecules with specific targets in redox signaling pathways, as shown in **Figure 1**.

Picture from "Commons Wikipedia" with license Creative Commons, created by Dan Cojocari. https://commons.wikimedia.org/w/index.php?curid=46529393

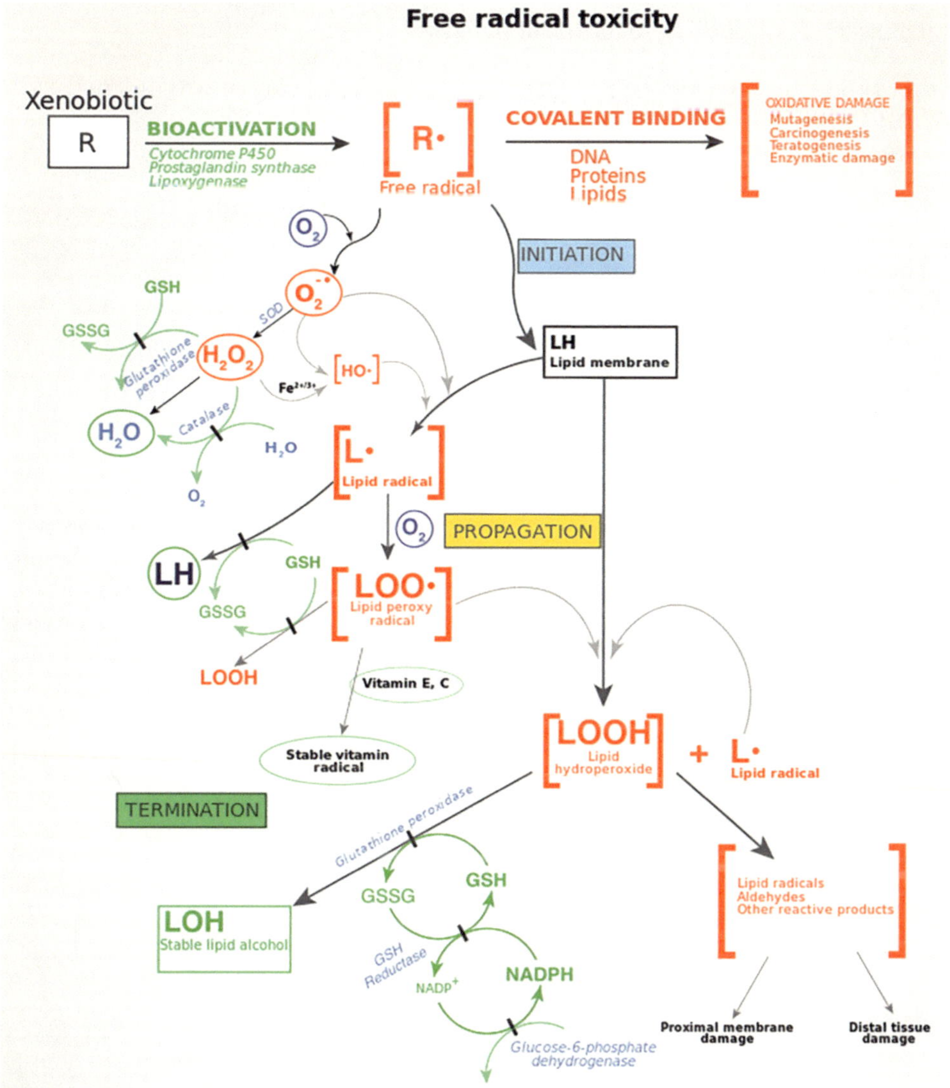

Figure 1.
Oxidative stress mechanisms in tissue injury. Free radical toxicity induced by xenobiotics and the subsequent detoxification by cellular enzymes.

RS and organism macromolecule bonding are controlled by the body's and external natural antioxidants, such as vitamins C, E, polyphenols, uric acid, various enzymes, and so on. Natural and artificial antioxidants can be got by common foods or food supplements (**Table 4**). There are two ways to distinguish the antioxidants [15]. The first is according to their function:

- Enzymatic antioxidants (superoxide dismutase, catalase [CAT], glutathione peroxidase (GPX)),
- Nonenzymatic antioxidants (glutathione (GSH), vitamins C, E, ubiquinones, carotenoids, β-carotene, lycopene, polyphenols, uric acid, and mineral antioxidants: zinc, copper, selenium).

The second is according to its line of defense:

- First line: the enzymes GSH, SOD, CAT,
- Second line: the vitamins A, C, D, E,
- Third line: carotenoids, the bioflavonoid, and coenzyme Q10.

The function of antioxidants can:

- Neutralize free radicals by exchanging electrons with them,
- Convert free radicals into other, less active RS (**Table 2**, Part 2),
- Act in combination, until they are completely inactivated [2].

At present, the role of all oxidants in human physiology and in diseases of various systems such as cardiovascular, immune system, skeletal muscles, metabolic regulation, aging, cancer, and reproduction is well established. Furthermore, oxidative stress is not the only cause of DNA damage. Long time before the role of RS in DNA damage was highlighted, the harmful effects of radiation and exogenous chemicals were already known. Once DNA damage is done by any cause, metabolic pathways are similar [16].

Substances	Food	Supplements
Vitamins C, E	Citrus fruits	Se
Flavonoids	Vegetables	Cu
Carotenoids	Red wine	Zn
Glutathione peroxidase	Green tea	Follic acid
Glutathione	Prunes	Q10
Ascorbic acid	Tomatoes	Vitamin C
Amino acids	Broccoli	Vitamin E
Uric acid	Olives, virgin olive oil	

Table 4.
Common antioxidants (in food and supplements).

2. The impact of oxidative stress on male infertility

The medical definition of "Infertility" is the failure to achieve pregnancy after 12 months of regular and unprotected sexual intercourse. It is a syndrome that affects not only a single person but also a couple. The male factor contributes to infertility in about 40–50% cases.

Although, high levels of oxidative stress increase male and female infertility, too, the actual data regarding only male infertility are presented here, since the oxidative stress effects on men have been studied in more detail than the corresponding effects on women, according to recent PubMed search. As it has been mentioned, RS contribute negatively to physiological processes, in cases when their concentration is found very high, thus disturbing the oxidant/antioxidant balance (**Tables 1, 2** and **4**) [17].

The following facts make oxidative stress particularly damaging to human sperm, which is known as more sensitive than other body cells:

- Oxidative stress damages the nucleus and mitochondria DNA. Spermatozoa contain much more mitochondria than other cells because of their high need for ATP necessary for their movement. For this reason, increased oxidative stress reduces sperm mobility.

- Spermatozoa are "special purpose" cells, since they have not developed all the cellular metabolic mechanisms, including antioxidant mechanisms. These mechanisms include the action of various enzymes such as superoxide dismutase (SOD), glutathione peroxidases (GPXs), peroxiredoxins (PRDXs), thioredoxins, and glutathione-S-transferases. In short, GPX4 is an essential component of the mitochondrial sheath in spermatozoa, while SOD2 protects the sperm genome during the maturation of the spermatozoon and controls mitochondrial superoxide (O_2^{-2}). On the other side, PRDXs reduce peroxide (H_2O_2) generated by SOD2 activity, preventing lipid peroxidation and DNA oxidation by scavenging H_2O_2 and $ONOO^-$ through its peroxidase activity and repairing oxidized membranes by its calcium-independent phospholipase A2 activity [18].

- Endogenous and exogenous mechanisms (**Tables 3, 4–6; Figure 2**), which alter the concentration of RS, can not only decrease the physical parameters of sperm (**Table 4**) but also affect their functionality (e.g., increase in apoptosis, increase in DNA fragmentation) [23].

RS are produced by all cells in the body, including sperm and white blood cells. However, in the restricted environment of male reproductive organs (testes, epididymis, vas deferens), there are few or many dead spermatozoa and many white blood cells in case of infections. Leukocytes produce up to 1000 times more RS than spermatozoa [24]. In **Figure 2**, stress mechanisms in male reproductive system and its consequences are presented.

In men, RS act detrimental to the seminal epithelium, causing the disease "idiopathic failure of the seminal epithelium," which is responsible for the 32% of male infertility, but its etiology generally remains still unclear. This disease is also known as idiopathic oligo-astheno-teratozoospermia (iOAT), since it produces (generate) a multi-decrease in sperm count (oligospermia), in sperm motility (asthenospermia), and in normal forms of spermatozoa (terato-zoospermia). The ROS impact on iOAT is

Internal sources/pathways & molecules	External sources
Cellular metabolism/Mitochondria	Exercise
Xanthine oxidase	Cigarette smoke
Peroxisomes	Environmental pollutants (drugs, pesticides, transition metals)
Phagocytosis	Food (lipids, carbohydrates, highly processed food)
Arachidonate pathways	Radiation
Drug metabolites	Industrial solvents
Internal sources/diseases	Ozone
Ischemia/reperfusion injury	Pathogens (bacteria, virus, fungus, parasite)
Anxiety	

Table 5.
Basic sources of RS production in humans [9].

Trace element	Normal daily intake for men	Dietary source
Zn	11 mg	Oyster, beef, poultry, seafood, cheese, legumes, grains
Cu	900 µg	Beef liver and shellfish (such as oysters), nuts (such as cashews), seeds (such as sesame and sunflower), chocolate, wheat-bran cereals, whole-grain products, potatoes, mushrooms, avocados, chickpeas, tofu
Se	55 µg	Seafood, meat, poultry eggs, breads, cereals, and other grain products

Table 6.
Basic sources of trace elements that influence oxidative stress and male infertility [19–22].

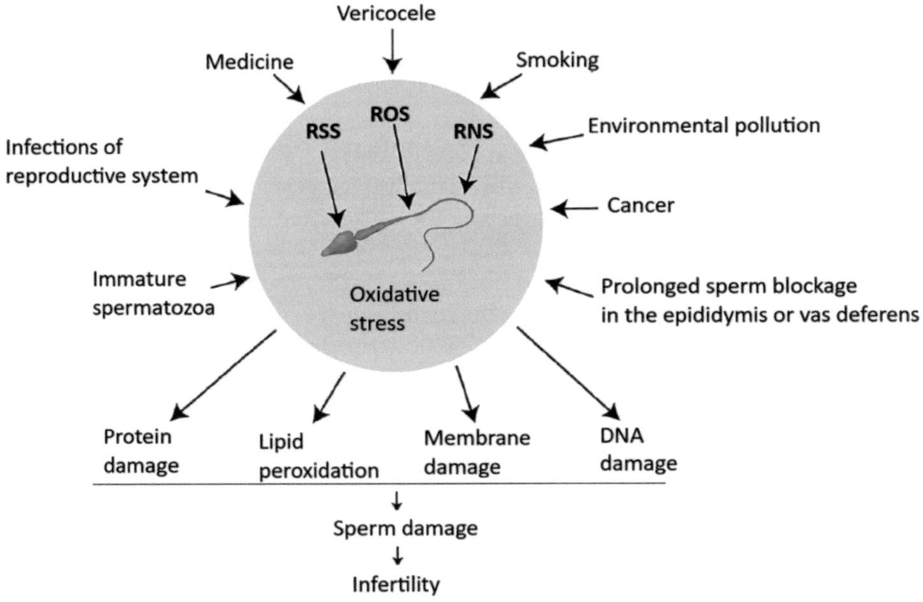

Figure 2.
Oxidative stress mechanisms in male reproductive system and its consequences.

confirmed by the improvement in spermatogenesis and when main parameter values of basic semen analysis (**Table 7**) after taking antioxidants (**Table 4**) or normal values of fertility hormones are well established (LH, FSH, PRL, Testosterone etc.) [26].

The adverse effects of oxidant-antioxidant imbalance can be obviously reduced by increasing antioxidants. The administration of antioxidants (see Supplements in **Table 4**) as dietary supplements may improve many sperm parameters [17].

2.1 Trace elements which influence oxidative stress

Trace elements are defined as specific inorganic elements involved in the body's metabolic processes in very small quantities. They are classified as essential and non-essential. Essential trace elements are absolutely necessary for the body's functions. Some of them (Ca, Na, K, Mg, Mn, Zn, Cu, Se) play an important role in human reproduction (**Figure 3**). Their deficiency reduces male fertility since it impairs the normal function of spermatogenesis (**Table 8**). The spermatogenesis decrease can be direct (deterioration of parameters of basic semen analysis) (**Table 7**) or indirect (decrease in steroid/testosterone production). Three of these trace elements affect male fertility by increasing or decreasing oxidative stress. These are selenium (Se), zinc (Zn), and copper (Cu) whose deficiency increases oxidative stress [25]. In **Figure 3**, **Tables 6** and **8**

Parameter	Reference values
Physical characteristics (7 parameters)	
pH	≥ 7.2
Odor	Normal
Consistency	Translucent
Color	Pale whitish or slightly yellowish
Volume	≥ 1.4 mL
Liquefaction	15–60 min
Viscosity	Normal
Microscopic characteristics (11 parameters)	
Agglutinates	No
Aggregates	Few, small
Sperm motility	≥ 32% Progressive motile spermatozoa ≥ 40% total motile spermatozoa
Vitality	≥ 58% livid spermatozoa
Sperm concentration	≥ 15,000.000/mL
Total sperm	≥ 39,000.000/ejaculate
Sperm morphology	≥ 4% normal forms
Round cells	< 5000.000/mL
White cells	< 1000.000/mL
Red cells	0–5 per field
Microorganisms	No

Table 7.
Measured parameters of basic semen analysis and their reference values [25].

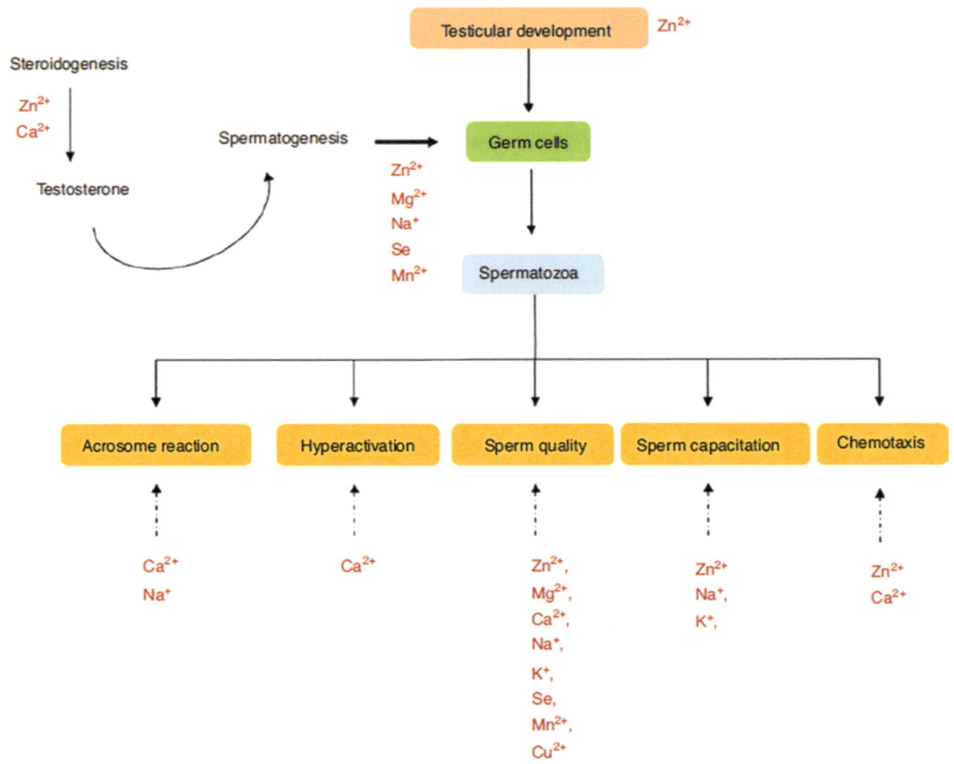

Figure 3.
The contribution of essential trace elements in male reproduction [25]. Under permission from the authors of the article "An overview on role of some trace elements in human reproductive health, sperm function and fertilization process. Reviews on Environmental Health."

offer details regarding chemical element contribution in male reproduction and sources of trace elements.

2.1.1 Role of Zn

It is the second most common trace element in the human body, after iron. Unlike iron, it is not stored in the body and must be obtained daily by diet (**Tables 6** and **9**). Although its value in human health is well documented and widely known, it is rarely measured in blood. Its determination in semen is more frequent and easier, since its concentration in seminal fluid is very high. The diagnostic significance of Zn determination in semen is particularly important, because Zn plays a major role in spermatogenesis and steroidogenesis (**Figure 3**; **Table 8**). It is present in high concentration in mature spermatozoa and spermatozoa's premature forms (spermatogonia and spermatids). Increase of seminal Zn increases male fertility, while seminal Zn decrease decreases male fertility.

The beneficial effects of Zn are (**Table 8**):

- Protects the testicles from heavy metals, temperature, and chlorine [27].
- It has a beneficial role in the function of the seminal epithelium and maintains normal levels of spermatogenesis [28]. Zn acts through proteins containing "*Zn*

Element	Place	Decrease	Increase
Ca deficiency	Seminal fluid	Steroidogenesis (testosterone) Sperm chemotaxis Sperm acrosome reaction Fertilization process Semen volume, sperm courts, sperm motility	—
Na & K deficiency	Seminal fluid	Fertilization rate Sperm quality Semen volume	—
Na deficiency	Seminal fluid	Progesterone Sperm acrosome reaction	—
	Intracellular	Semen capacitation	—
K deficiency	Seminal fluid	Testosterone	—
Mg deficiency	Seminal fluid	Premature ejaculation Sperm motility	
Zn deficiency	Seminal fluid	Sperm quality Testicular development & function Sexual maturation Steroidogenesis (testosterone) Spermatogenesis Sperm quality	Hypogonadism Gonad dysfunction Testicular weight Leydig cell damage Sperm quality Lipid peroxidation Sperm membrane fluidity Oxidative stress
Se deficiency	Dietary intake	Spermatogenesis Sperm quality	Oxidative stress
Se deficiency	Seminal fluid	Spermatogenesis Secretion of testosterone Sperm count Motility Normal morphology Vitality	—
Mn deficiency	Seminal fluid	Sperm morphology Seminal fluid volume	—
Mn increase	Seminal fluid	Sperm motility Sperm count	—
Cu deficiency	Seminal fluid	Sperm quality	Oxidative stress
Cu increase	Seminal fluid	Sperm motility	—

The decrease of Se, Zn, and Cu increase oxidative stress.

Table 8.
Trace elements that influence men infertility [25].

finger" structures (Cys2/His2 P2 protamine). Its deficiency can cause oligo-asthenospermia (low motility, number, and normal concentration of spermatozoa) or even azoospermia (complete lack of spermatozoa).

- Possesses antioxidant activity. Proteins containing Zn and Se bind large amounts of ROS and reduce the effects of oxidative stress. Zn is a necessary component of the enzyme Cu/Zn superoxide dismutase, which demonstrates antioxidative activity for sperm function and inhibits DNA fragmentation [29]. Low Zn

Food	mg per serving	Daily Value (%)*
Oysters, Eastern, farmed, raw, 3 ounces	32	291
Oysters, Pacific, cooked, 3 ounces	28.2	256
Beef, bottom sirloin, roasted, 3 ounces	3.8	35
Blue crab, cooked, 3 ounces	3.2	29
Breakfast cereals, fortified with 25% of the DV for zinc, 1 serving	2.8	25
Cereals, oats, regular and quick, unenriched, cooked with water, 1 cup	2.3	21
Pumpkin seeds, roasted, 1 ounce	2.2	20
Pork, center loin (chops), bone in, broiled, 3 ounces	1.9	17
Turkey breast, meat only, roasted, 3 ounces, Cheese cheddar, 1.4 ounces	1.5	14
Shrimp, cooked, 3 ounces	1.4	13
Lentils, boiled, ½ cup	1.3	12
Sardines, canned in oil, drained solids with bone, 3 ounces	1.1	10
Greek yogurt, plain, 6 ounces	1.0	9

**Daily value for Zinc is 11 mg.*

Table 9.
Quantity of Zn in the most richly food [20].

seminal levels are accompanied by male infertility especially after smoking and alcohol consumption.

- It has as antibacterial activity against Gram-negative and Gram-positive bacteria and trichomonas vaginalis [30].

- Increases the vitality of spermatozoa and shows anti-lipid peroxidation properties that maintain membrane stability of spermatozoa and other testis cells, that is, Sertoli, Leybig [31]. In addition, its antioxidant properties reduce the activity of DNases and reduce the concentration of RS produced by leukocytes (in case of infections) and dead spermatozoa [32, 33].

- Regulates the reproductive activity of spermatozoa. It has a regulative role in the progress of capacitation of sperm inside female reproductive system and acrosome reaction (connection of sperm with oocyte) [34].

- Regulates the production and function of various cells of the immune system. Zn regulates the production, maturation, and function of leucocytes (PMN-cells, B-cells, NK cells, Pre-T cells, Monocytes, T helper cells), hence influencing the function of immunostimulants used in the experimental systems [35, 36]. It probably influences the anti-sperm antibody production.

On the contrary, the lack of Zn causes:

- All the expected damage to cells, caused by oxidative stress, such as DNA damage and apoptosis.

- Decrease of spermatozoa's motility, number, and normal morphology. Sperm motility shows the largest decrease, since Zn is present in high concentration in their tail and influences its normal function.

- Decrease the gonadotropin-releasing hormone (GnRH), which stimulates the production of gonadotropins (FSH, LH) [37]. The decrease of FSH reduces testosterone production by *Leybig* cells and spermatogenesis, which is controlled by *Sertoli* cells under the influence of testosterone.

In seminal fluid, Zn levels can be measured by photometric method [38] and flame atomic absorption [39]. Due to its high concentration in seminal fluid, it is also a biomarker of the prostate, since its high decrease can be attributed to prostate anatomical damage.

2.1.2 Role of Se

Selenium is required for normal spermatogenesis (**Table 8**) since its deficiency (<50 μg/day, **Table 6**) reduces sperm motility, vitality, total number, and morphology (**Table 7**). It is essential for the action of selenoproteins (PHGPx) found on the cells of male gametes. Selenium deficiency is very rare. Except male infertility Se deficiency can cause *Keshan* disease (a type of heart disease) and *Kashin-Beck* disease, a type of arthritis that produces pain, swelling, and loss of motion in joints [40]. Selenium may have a positive influence on *Leydig* cells, thus influencing the secretion of testosterone and consequently spermatogenesis.

The measurement of Se levels in clinical laboratory is done by flame atomic absorption [41].

2.1.3 Role of Cu

Cu is an essential element for numerous metalloproteins, including Cu/Zn-SOD, cytochrome C, oxidase, and tyrosinase, which are involved in energy and antioxidant metabolism. It also contributes to redox system and hence protects sperm cells against oxidative damages (**Table 6**). Superoxide dismutase (SOD) catalyzes the dismutation (or partitioning) of the superoxide anion radical into normal molecular oxygen and hydrogen peroxide [42].

Copper as a heavy metal can be accumulated in the body from nature or industrial pollution, hence it can become toxic at high concentrations (**Table 6**) to many body systems including the reproductive system [25]. At concentrations above 19 μg/mL, it limits the activity of sperm mitochondria and thus their mobility [42]. At concentrations above 100 μg/mL, it has a negative effect not only on mobility but also on sperm morphology [43] and DNA [44]. The lab determination of Cu requires special equipment and is done by flame atomic absorption [39].

3. Determination methods of total oxidative capacity in seminal fluid

Total oxidative capacity (TOC) is the total concentration of all RS that create oxidative stress in biological fluids. TOC can be determined in seminal fluid either directly or indirectly. The direct measurement involves the measurement of RS with several methodologies, some of which are commercially available. Indirect determination can be done by measurement trace elements that affect the concentration of oxidative stress (Zn, Se, Cu) or by estimating the impact of TOC on male fertility, such as:

- Parameters of basic semen analysis (**Table 4**),
- DNA fragmentation index,
- Apoptosis of spermatozoa.

3.1 Measurement of oxidative stress in seminal fluid with commercial kits

One of the commercial methods to determine oxidative stress in semen is the Oxisperm method. The Oxisperm kit is produced by Halotech DNA and has IVD certification according to EC Directive 746/2017.

The method is based on the ability of water-soluble nitro blue tetrazolium (NBT) to be converted under the influence of peroxide anions (O_2^{-2}) of oxidative stress into water-soluble blue crystals, known as formazan. Formazan crystals are visible in the cell membranes of spermatozoa. They can be observed by microscopy at 400X magnification under a bright field of view [45]. In addition, formazan crystals produce a characteristic color effect when reacted with agarose gel (Reactive Gel), which varies from yellow to purple-blue. The intensity of the color is proportional to the level of oxidative stress (**Figure 4**). The assessment of the color intensity can be done either visually or by measuring the absorbance of the color in a photometer at a wavelength of 400–600 nm.

A different approach to assessing oxidative stress has been proposed by BRED Life Science Technology, which assesses the effect of ROS on sperm through the effect of superoxide anions on sperm mitochondria. According to this method, the mitochondria of living spermatozoa that have been exposed to the effect of superoxide anions

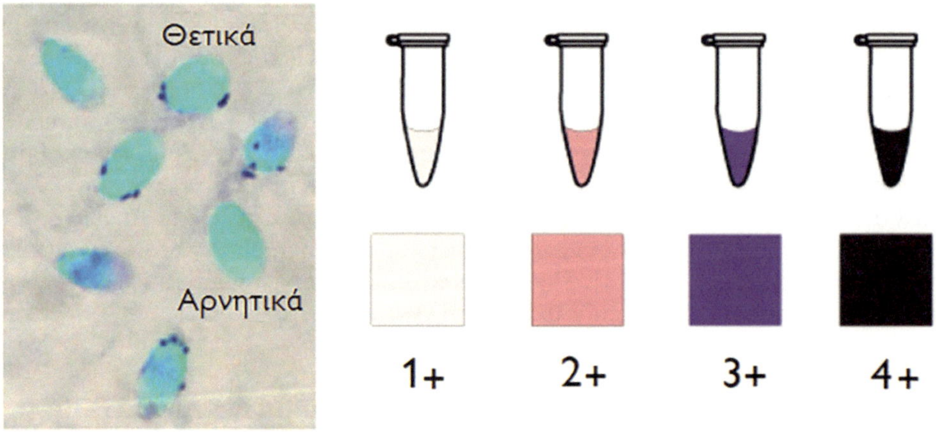

Figure 4.
On the left side, positive and negative sperm heads that have reacted (positive formosan) or not (negative formosan) with NBT. NBT gives a positive reaction in sperm heads where it reacts with hydroxyl anions which are inside eppendorf tube with agarose gel. The intensity of the color corresponds to the intensity of the reaction. Under permission from "Helix2 Inc."

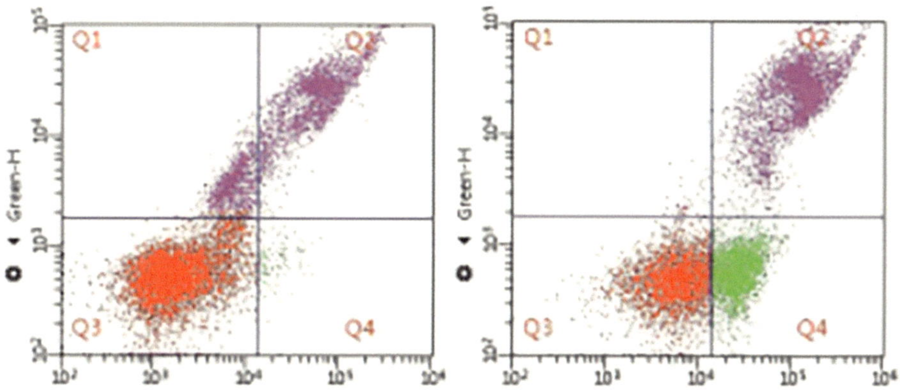

Figure 5.
Two different cytometry images showing a negative (left) and a positive (right) semen sample under the influence of superoxide anions on the mitochondria of live spermatozoa. The four quadrants of the cytometric nephelogram depict the following: Q1: dead spermatozoa, O2⁻ mitochondria negative; Q2: dead spermatozoa, O_2- mitochondria positive; Q3: live spermatozoa, O2- mitochondria (negative); Q4: live spermatozoa, O_2- mitochondria (positive). Under permission from "BRED Life Science Technology Inc."

are labeled differently from the rest of the spermatozoa (**Figure 5**). The result can be seen by flow cytometry. The advantage of this method is that the determination of oxidative stress is carried out in living spermatozoa. However, special analytical equipment (flow cytometry analyzer) is required [46].

3.1.1 Determination of oxidizing capacity by FRAP method

FRAP method is based on measuring the ability of sperm antioxidants to reduce the Fe^{3+}-TPTZ complex to Fe^{2+}-TPTZ. TPTZ is the substance (2,4,6-tripyridyl-s-triazine). The reduction takes place at low pH and gives a solution of intense blue color, the absorbance of which is measured at a wavelength of 593 nm. The reaction is non-specific since any semi-conducting reaction with a lower redox potential than the trivalent-divalent iron reaction under the reducing conditions of the reaction will result in the formation of Fe^{2+} ions. The absorbance change is directly related to the "total" reducing capacity of the antioxidants (as electron donors) inside the reaction mixture [47].

3.1.2 Determination of oxidizing capacity by the MDA method

Malondialdehyde (MDA) is a low molecular-weight molecule that is produced by the decomposition of the peroxides. MDA is considered the best biomarker of lipid peroxidation, which is one of effects of RS on membrane cells. The produced MDA reacts with TBA, forming an MDA-TBA2 compound that absorbs strongly at 532 nm [48]. There are MDA IVD kits in the trade like TBARS & NWLSSTM assays.

Because most of peroxides are produced by multi-unsaturated lipids (PUFA), MDA measures mostly them [49]. PUFA, which have two or more double bonds, are easily associated with free radicals of various RS (i.e., hydroxyl radicals [HO^-]). The connection of PUFA with HO^- produces a lipid peroxyl radical (LOO^-) that reacts with a second PUFA and produces a lipid hydroperoxide (LOOH) and a second peroxyl radical (LOO^-). LOO^- forms an intramolecular double bond, and it can be transformed to a cyclic endoperoxide [50].

3.2 Determination of DNA fragmentation index

DNA Fragmentation Index (DFI) expresses the percentage of spermatozoa that retain their DNA intact. In healthy men, there is a percentage of spermatozoa (<15%) where their DNA is damaged even though the spermatozoa retain good morphology and/or other microscopic characteristics. Oxidative stress increases DFI % by more than 15%. The high values of DFI decrease male fertility.

The reasons of high values of RS are the main indications for the measurement of DFI in infertile men, who have/are:

- history of multiple miscarriages with their partner,
- unexplained infertility for more than 6 months from both male and female factors,
- over 40 years old,
- smoking habits,
- obesity,
- history of cancer,
- medications for various reasons,
- exposed to toxic agents or long-term exposure to high temperatures,
- urogenital infection,
- failed to develop healthy embryos in IVF cycles.

A common method for the determination of DFI is Sperm Chromatin Dispersion (SCD) [51, 52]. The principle of the method includes the fixing of spermatozoa in an agarose gel and then on a slide. The preparation is then treated with a denaturing agent to decompose the fragmented DNA. Lysis solution is then added to remove the cytoplasm of the spermatozoa. The remaining nuclear material is then stained. The level of DNA fragmentation is determined by the halo observed at the head of the spermatozoa. Spermatozoa with high DNA fragmentation have no halo or minimal halo, whereas normal spermatozoa release DNA chains and create a moderate to large diameter halo (**Figure 6**) [53].

3.3 The determination of sperm apoptosis

Apoptosis (cell death) is another consequence of oxidative stress. A common method to determine apoptosis is TUNEL [38]. TUNEL stands for "terminal deoxynucleotidyl transferase dUTP nick end labeling." In the TUNEL method, fragments of sperm DNA are detected after being initially labeled with deoxyuridine triphosphate (dUTP). This is followed by binding to the final tracer, which can be either fluorochrome (**Figure 7**) or biotin-streptavidin-peroxidase (HRP) complexes that react

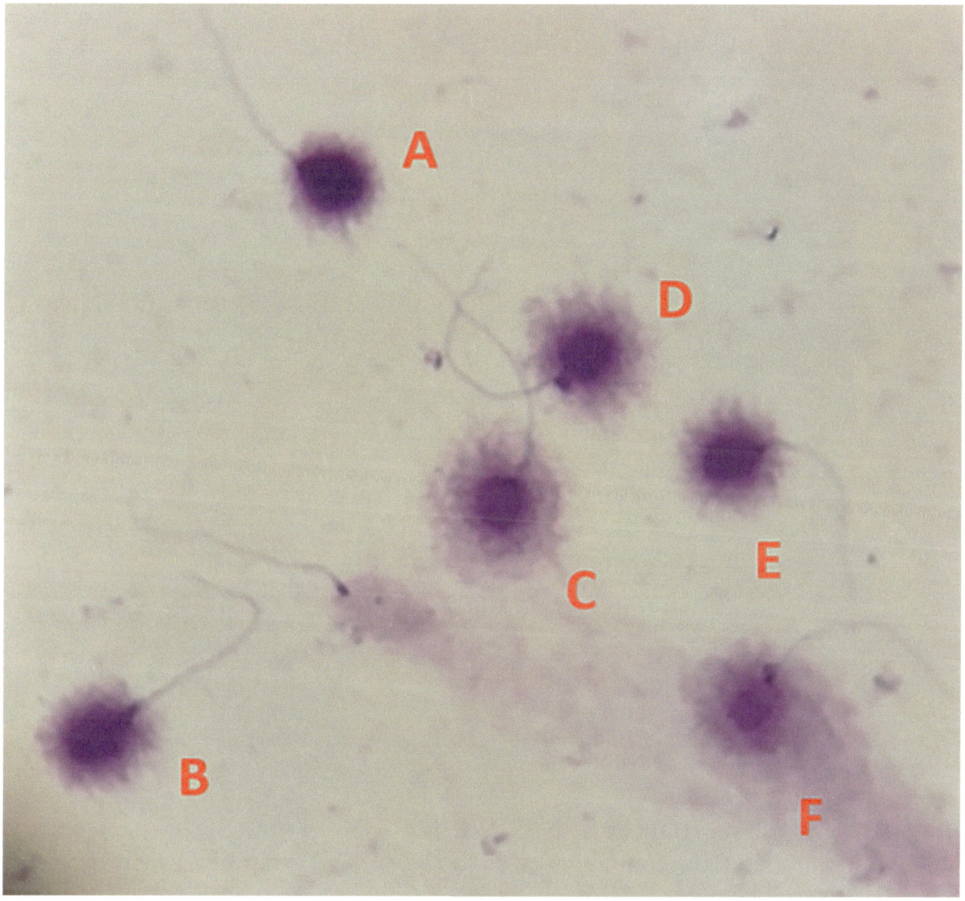

Figure 6.
Among the stained spermatozoa of the picture, spermatozoon A has a small halo, and it is considered abnormal; the spermatozoa B, D, and E have medium halo, and C and E have large halo. Spermatozoa with medium halo (B, D, E) and large halo (C, E) are considered normal. The photo comes from the personal archive of the authors.

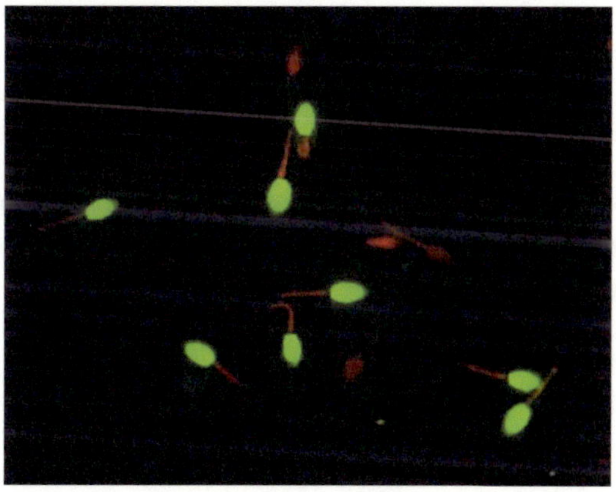

Figure 7.
Positive apoptotic spermatozoa (green) and negative apoptotic spermatozoa (red) by fluorescence microscope. The photo comes from the personal archive of the authors.

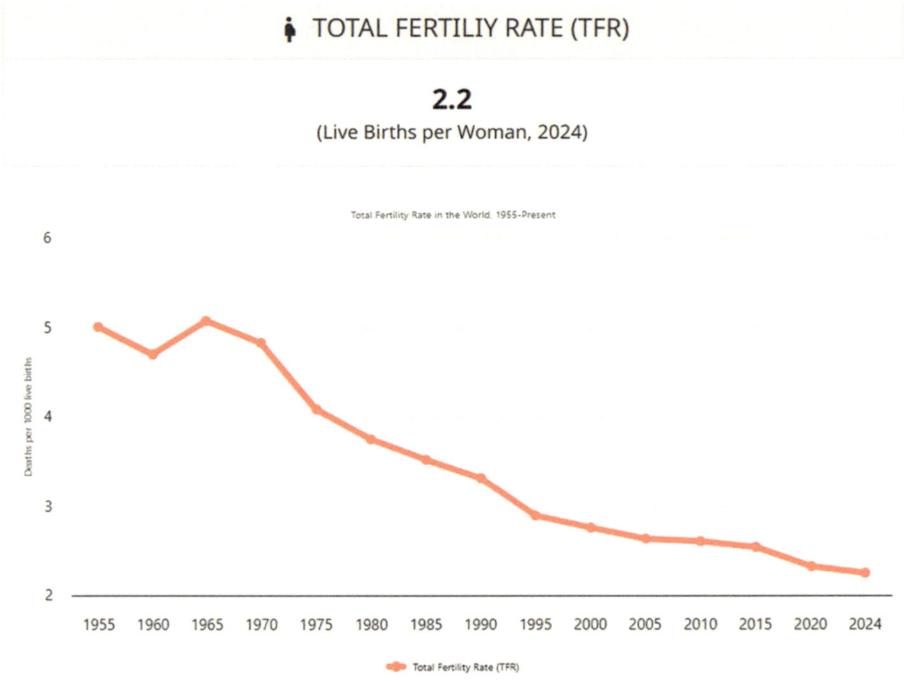

Figure 8.
The decreasing of Total Fertility Rate (TRF) from 1955 (5 births per woman) until 2014 (2.2 births per woman). Free picture from UN Worldmeter (https://www.worldometers.info/).

with a chromogenic substrate of HRP. TUNEL is capable of directly assessing both single-strand and double-strand breaks, so the more DNA strand fragments present, the greater the cell labeling [54].

4. Conclusions

Total Fertility Rate has been decreasing since 1955 (**Figure 8**). That year is considered the highlight of "baby boom," in births, after the Second World War. The reasons of today's birth decline are social (e.g., urbanization, changing social status of women), economic (e.g., consumerism, delay in economic independence), and most importantly, the current unhealthy lifestyle habits (e.g., bad diet, smoking and alcohol consumption, no body exercise, inadequate sleep, no regular daily activities, etc.). Many health problems, due to the above general homeostasis disturbances, influence among other the reproductive ability of both men and women. The increase of oxidative stress is considered the "silent" reason of human infertility. Oxidative stress is the chemical explanation for the effect of today's urban lifestyle in reducing human fertility, and this condition is not only due to oxygen species since nitrogen and sulfur ions and their neutral compounds can also demonstrate similar chemical effects as oxygen ions in cell metabolism (**Table 1**).

In conclusion, the effects of oxidative stress can be decreased by reducing the RS production (**Table 5**) or/and by taking antioxidant substances (**Table 4**). The confirmed antioxidant activity of three common trace elements (Zn, Se, Cu) of human metabolism reduces the effects of oxidative stress on spermatozoa. The most important for fertility is Zn, produced by the prostate gland. Its concentration can be increased exogenously through consumption of certain foods rich in Zn (**Table 9**). The relative ratio of Zn to other seminal substances, for example, fructose, indicates the ratio of prostatic fluid to seminal vesicles' fluid. Therefore, a high amount of prostatic fluid in the semen is a good "bio-marker" for the best male quality sperm.

Antioxidant diet has been used for the treatment of male infertility [54, 55]. **Table 10** contains the proposed antioxidants and their doses for the treatment of

Diagnosis	Antioxidant
Oligozoospemia	2000 μg of Lycopene twice a day
	N-acetyl cysteine 10 mg/kg/diet, Vitamin C 3 mg/kg/diet, Vitamin E 0.2 mg/kg/diet, Vitamin A 0.06 IU/kg/diet, Thiamine 0.4 mg/kg/diet, Riboxavin 0.1 mg/kg/diet, piridoxin 0.2 mg/kg/diet, Nicotinamide 1 mg/kg/diet, pantothenate 0.2 mg/kg/diet, Biotin 0.04 mg/kg/diet, Cyanocobalamin 0.1 mg/kg/diet, Ergocalciferol 8 IU/kg/diet, Calcium 1 mg/kg/diet, Magnesium 0.35 mg/kg/diet, Phosphate 0.45 mg/kg/diet, iron 0.2 mg/kg/diet, Manganese 0.01 mg/kg/diet, Copper 0.02 mg/kg/diet, Zinc 0.01 mg/kg/diet
	L-carnitine (2 g)
	CoQ10 (300 mg)
	200 μg Selenium orally daily, 600 mg N-acetyl-cysteine orally daily, 200 μg Selenium plus 600 mg N-acetyl-cysteine orally daily
	CoQ10 (200 mg)
	Folic acid (5 mg) and Zinc (66 mg)
	Zinc sulphate 200 mg twice daily, Zinc sulphate 200 mg + Vitamin E 10 mg twice daily, Zinc sulphate 200 mg + Vitamin E 10 mg + Vitamin C 5 mg twice daily
	N-acetyl cysteine (600 mg) and Selenium (200 mg)
	Lycopene (2 mg)
	Selenium (200 mug) in combination with Vitamin E (400 units)
	L-carnitine 2 g/day and L-acetyl cysteine 1 g/day
	L-carnitine 3 g/day or/and L-acetyl cysteine 3 g/day
	L-carnitine (2 g/day) + L-acetyl cysteine (1 g/day) + Cinnoxicam 30 mg/day
Asthenozoospermia	N-acetyl cysteine (600 mg) and selenium (200 mg)
	Lycopene (2 mg)
	1 g vitamin C and 1 g Vitamin E
	N-acetyl cysteine (600 mg)
	CoQ10 (300 mg)
	CoQ10 (200 mg)
	l-carnitine (2 g/day) and l-acetyl-carnitine (1 g/day)
	L-carnitine (2 g) and L-acetyl cysteine (1 g)
	N-acetyl cysteine (600 mg/day orally)
	L-carnitine 3 g/day and L-acetyl cysteine 3 g/day
	Zinc 500 mg/day
	L-carnitine (2 g/day) + L-acetyl cysteine (1 g/day) + Cinnoxicam 30 mg/day

Diagnosis	Antioxidant
Teratozoospermia	L-carnitine (2 g/day) and L-acetyl-carnitine (1 g/day)
	N-acetyl cysteine (600 mg) and Selenium (200 mg)
	1 g Vitamin C and 1 g Vitamin E
	CoQ10 (200 mg)
	Lycopene (2 mg)
	Vitamins C and E (400 mg each), β-carotene (18 mg), Zinc (500 μmol) and Selenium (1 μmol)
	L-carnitine 3 g/day and L-acetyl cysteine 3 g/day
	L-carnitine (2 g/day) + L-acetyl cysteine (1 g/day) + Cinnoxicam (NSAID) 30 mg/day
High oxidative stress	2000 μg of Lycopene
	l-carnitine (2 g/day) and l-acetyl-carnitine (1 g/day)
	Vitamin E (400 mg) and Selenium (225 g)
	Lycopene 6 mg Vitamin E 400 IU Vitamin C 100 mg Zinc 25 mg Selenium 26 μgm Folate 0.5 mg Garlic 1000 mg
	N-acetyl cysteine (600 mg)
	1 g Vitamin C and 1 g Vitamin E
	Vitamin E (600 mg)
	Vitamin E (400 mg) and Selenium (225 g)
High DNA fragmentation	Vitamin C (400 mg), Vitamin E (400 mg), b-carotene (18 mg), Zinc (500 mmol) and Selenium (1 mmol)
	L-carnitine (1500 mg); vitamin C (60 mg); CoQ10 (20 mg); vitamin E (10 mg); zinc (10 mg); folic acid (200 μg), selenium (50 μg); vitamin B12 (1 μg)
	Vitamin E 100 mg
Improving sperm function tests	Vitamin E (600 mg/day)
	Vitamin E 100 mg
	L-carnitine/L-acetyl-carnitine +1 × 30-mg cinnoxicam, L-carnitine (2 g/day) + L-acetyl-carnitine (1 g/day)
	Vitamin E (400 mg) and Selenium (225 g)
	Vitamin E 100 mg
	Vitamin C (1 g) + vitamin E (1 g)

Table 10.
The proposed antioxidants and their doses for the treatment of specific sperm abnormal parameters [56].

specific pathological sperm parameters (oligozoospermia, teratozoospermia, asthenospermia, high oxidative stress, high sperm defragmentation, etc.).

Funding

The project is financially supported by ELKE UniWA (Special Account for Research Funds of University of West Attica).

DOI: http://dx.doi.org/10.5772/intechopen.1008873

Abbreviations

CAT	catalase
Cu	copper
DA	denaturant agent
DFI	DNA fragmentation index
DNA	deoxyribonucleic acid
dUTP	deoxy uridine triphosphate
EC	European community
FR	free radicals
FRAP	fluorescence recovery after photobleaching
FSH	follicle-stimulating hormone
GnRH	gonadotropin-releasing hormone
GPx	glutathione peroxidase
GSH	glutathione
HRP	horse raddish peroxidase
iOAT	idiopathic oligo αstheno τeratozoospermia
IVD	*in vitro* diagnostics
K	potassium
LH	luteinizing hormone
MDA	malondialdehyde
Mg	magnesium
Mn	µanganese
Na	sodium
NBT	nitro blue tetrazolium
PHGPx	phospholipid hydroperoxide glutathione peroxidase
PUFA	poly unsaturated fatty acids
RG	reactive gel
RNA	ribonucleic acid
RNS	reactive nitrogen species
ROS	reactive oxygen species
RS	reactive species
RSS	reactive sulfur species
SCD	sperm chromatin dispersion
Se	selenium
SOD	superoxide dismutase
TAC	total antioxidant capacity
TBA	τhiobarbituric acid
TBARS	thiobarbituric acid reactive substances
TOC	total oxidative capacity
TPTZ	2,4,6-tripyridyl-s-triazine
TUNEL	terminal deoxynucleotidyl transferase dUTP nick end labeling
Zn	zinc

Author details

Petros Karkalousos*, Maria Trapali and George Albert Karikas
Laboratory of Chemistry, Biochemistry and Cosmetic Science, Department of Biomedical Sciences, University of West Attica, Athens, Greece

*Address all correspondence to: petef@uniwa.gr

References

[1] Rice-Evans CA, Gopinathan V. Oxygen toxicity, free radicals and antioxidants in human disease: Biochemical implications in atherosclerosis and the problems of premature neonates. Essays in Biochemistry. 1995;**29**:39-63

[2] Lü JM, Lin PH, Yao Q, Chen C. Chemical and molecular mechanisms of antioxidants: Experimental approaches and model systems. Journal of Cellular and Molecular Medicine. 2010;**14**(4):840-860. DOI: 10.1111/j.1582-4934.2009.00897.x

[3] Radi R. Peroxynitrite, a stealthy biological oxidant. The Journal of Biological Chemistry. 2013;**288**(37):26464-26472. DOI: 10.1074/jbc.R113.472936

[4] Giles GI, Nasim MJ, Ali W, Jacob C. The reactive sulfur species concept: 15 years on. Antioxidants. 2017;**6**(2):38. DOI: 10.3390/antiox6020038

[5] Zaha I, Muresan M, Tulcan C, Huniadi A, Naghi P, Sandor M, et al. The role of oxidative stress in infertility. Journal of Personalized Medicine. 2023;**13**:1264-1277. DOI: 10.3390/jpm13081264

[6] Vona R, Pallotta L, Cappelletti M, Severi C, Matarrese P. The impact of oxidative stress in human pathology: Focus on gastrointestinal disorders. Antioxidants (Basel). 2021;**10**(2):201-217. DOI: 10.3390/antiox10020201

[7] Burton GJ, Jauniaux E. Oxidative stress. Best Practice & Research. Clinical Obstetrics & Gynaecology. 2011;**25**(3):287-299. DOI: 10.1016/j.bpobgyn.2010.10.016

[8] Yang S, Lian G. ROS and diseases: Role in metabolism and energy supply. Molecular and Cellular Biochemistry. 2020;**467**(1-2):1-12. DOI: 10.1007/s11010-019-03667-9

[9] Lobo V, Patil A, Phatak A, Chandra N. Free radicals, antioxidants and functional foods: Impact on human health. Pharmacognosy Reviews. 2010;**4**(8):118-126. DOI: 10.4103/0973-7847.70902

[10] Gomberg M. An instance of trivalent carbon: Triphenylmethyl. Journal of the American Chemical Society. 1900;**22**:757-771. DOI: 10.1021/ja02049a006

[11] Michaelis L. Free radicals as intermediate steps of oxidation-reduction. Cold Spring Harbor Symposia on Quantitative Biology. 1939;**7**:33-49. DOI: 10.1101/SQB.1939.007.01.003

[12] Commoner B, Townsend J, Pake GE. Free radicals in biological materials. Nature. 1954;**174**:689-691. DOI: org/10.1038/174689a0

[13] Sies H. Oxidative stress: Introductory remarks. In: Sies H, editor. Oxidative Stress. London: Academic Press; 1985. pp. 1-8

[14] Egbuna C. Antioxidants and phytochemicals. In: Phytochemistry. USA: Apple Academic Press; 2018. p. 16. eBook. ISBN9780429426193

[15] Sies H, Jones DP. Reactive oxygen species (ROS) as pleiotropic physiological signaling agents. Nature Reviews. Molecular Cell Biology. 2020;**21**:363-383. DOI: 10.1038/s41580-020-0230-3

[16] Sadasivam N, Kim YJ, Radhakrishnan K, Kim DK. Oxidative stress, genomic integrity, and liver diseases. Molecules. 2022;**27**(10):3159. DOI: 10.3390/molecules27103159

[17] Takalani NB, Monageng EM, Mohlala K, Monsees TK, Henkel R, Opuwari CS. Role of oxidative stress in male infertility. Reproduction and Fertility. 2023;**4**(3):e230024. DOI: 10.1530/RAF-23-0024

[18] O'Flaherty C, Scarlata E. Oxidative stress and reproductive function: The protection of mammalian spermatozoa against oxidative stress. Reproduction. 2022;**164**(6):F67-F78. DOI: 10.1530/REP-22-0200

[19] Sangeetha VJ, Dutta S, Moses JA, Anandharamakrishnan C. Zinc nutrition and human health: Overview and implications. eFood. 2022;**3**:e17. DOI: 10.1002/efd2.17

[20] NIH. Available from: https://ods.od.nih.gov/factsheets/Zinc-HealthProfessional/#en11 [Accessed: August 18, 2024]

[21] NIH. Available from: https://ods.od.nih.gov/factsheets/Selenium-Consumer/ [Accessed: August 18, 2024]

[22] NIH. Available from: https://ods.od.nih.gov/factsheets/Copper-Consumer [Accessed: August 18, 2024]

[23] Bisht S, Faiq M, Tolahunase M, Dada R. Oxidative stress and male infertility. Nature Reviews. Urology. 2017;**14**(8):470-485. DOI: 10.1038/nrurol.2017.69

[24] Alahmar AT. Role of oxidative stress in male infertility: An updated review. Journal of Human Reproductive Sciences. 2019;**12**:4-18. DOI: 10.4103/jhrs.JHRS_150_18

[25] Mirnamniha M, Faroughi F, Tahmasbpour E, Ebrahimi P, Beigi Harchegani A. An overview on role of some trace elements in human reproductive health, sperm function and fertilization process. Reviews on Environmental Health. 2019;**34**:339-348. DOI: 10.1515/reveh-2019-0008

[26] Agarwal A, Sekhon L. Oxidative stress and antioxidants for idiopathic oligoasthenoteratospermia: Is it justified? Indian Journal of Urology. 2011;**27**:74-81. DOI: 10.4103/0970-1591.78437

[27] Boran C, Ozkan KU. The effect of zinc therapy on damaged testis in pre-pubertal rats. Pediatric Surgery International. 2004;**20**(6):444-448. DOI: 10.1007/s00383-004-1173-z

[28] Colagar AH, Marzony ET, Chaichi MJ. Zinc levels in seminal plasma are associated with sperm quality in fertile and infertile men. Nutrition Research. 2009;**29**(2):82-88. DOI: 10.1016/j.nutres.2008.11.007

[29] Fallah A, Mohammad-Hasani A, Colagar AH. Zinc is an essential element for male fertility: A review of Zn roles in men's health, germination, sperm quality, and fertilization. Journal of Reproduction & Infertility. 2018;**19**(2):69-81

[30] Vijayalakshmi K, Sivaraj D. Enhanced antibacterial activity of Cr doped ZnO nanorods synthesized using microwave processing. RSC Advances. 2015;**5**(84):68461-68469. DOI: 10.1039/C5RA13375K

[31] Eggert-Kruse W, Zwick EM, Batschulat K, Rohr G, Armbruster FP, Petzoldt D, et al. Are zinc levels in seminal plasma associated with seminal leukocytes and other determinants of semen quality? Fertility and Sterility. 2002;**77**(2):260-269. DOI: 10.1016/s0015-0282(01)02974-0

[32] Ebisch IM, Thomas CM, Peters WH, Braat DD, Steegers-Theunissen RP. The importance of folate, zinc and

antioxidants in the pathogenesis and prevention of subfertility. Human Reproduction Update. 2007;**13**(2):163-174. DOI: 10.1093/humupd/dml054

[33] Josarayi GA, Mohammad-Hasani A, Aftabi Y, Moudi E, Hosseinzadeh CA, The AhRR-c. 565C> G transversion may increase total antioxidant capacity levels of the seminal plasma in infertile men. Environmental Science and Pollution Research International. 2017;**24**(21):17428-17435. DOI: 10.1007/s11356-017-9356-z

[34] Michailov Y, Ickowicz D, Breitbart H. Zn^{2+}-stimulation of sperm capacitation and of the acrosome reaction is mediated by EGFR activation. Developmental Biology. 2014;**396**(2):246-255. DOI: 10.1016/j.ydbio.2014.10.009

[35] Rink L, Gabriel P. Zinc and the immune system. The Proceedings of the Nutrition Society. 2000;**59**(4):541-552. DOI: 10.1017/s0029665100000781

[36] Wessels I, Maywald M, Rink L. Zinc as a gatekeeper of immune function. Nutrients. 2017;**9**(12):1286. DOI: 10.3390/nu9121286

[37] Hunt CD, Johnson PE, Herbel J, Mullen LK. Effects of dietary zinc depletion on seminal volume and zinc loss, serum testosterone concentrations, and sperm morphology in young men. The American Journal of Clinical Nutrition. 1992;**56**(1):148-157. DOI: 10.1093/ajcn/56.1.148

[38] WHO. Laboratory Manual for the Examination and Processing of Human Sperm. WHO Editions. 6th ed. Geneva, Switzerland: World Health Organization; 2021

[39] Afridi I, Gul Kazi T, Naz Talpur F, Ahmed Baig J, Qadir CG. Essential trace and toxic elemental concentrations in biological samples of male adult referent and eunuch subjects. Clinica Chimica Acta. 2022;**529**:96-103. DOI: 10.1016/j.cca.2022.02.010

[40] Akinloye O, Arowololu AO, Shittu OB, Adejuwon CA, Osotimehin B. Selenium status of idiopathic infertile Nigerian males. Biological Trace Element Research. 2005;**104**:9-18. DOI: 10.1385/BTER:104:1:009

[41] Morabbi A, Karimian M. Trace and essential elements as vital components to improve the performance of the male reproductive system: Implications in cell signaling pathways. Journal of Trace Elements in Medicine and Biology. 2024;**83**:127403. DOI: 10.1016/j.jtemb.2024.127403

[42] Knazicka Z, Tvrda E, Bardos L, Lukac N. Dose- and time-dependent effect of copper ions on the viability of bull spermatozoa in different media. Journal of Environmental Science and Health. Part A, Toxic/Hazardous Substances & Environmental Engineering. 2012;**47**:1294-1230. DOI: 10.1080/10934529.2012.672135

[43] Roblero L, Guadarrama A, Lopez T, Zegers-Hochschild F. Effect of copper ion on the motility, viability, acrosome reaction and fertilizing capacity of human spermatozoa *in vitro*. Reproduction, Fertility and Development. 1996;**8**:871-874. DOI: 10.1071/rd9960871

[44] Wang YX, Wang P, Feng W, Liu C, Yang P, Chen YJ, et al. Relationships between seminal plasma metals/metalloids and semen quality, sperm apoptosis and DNA integrity. Environmental Pollution. 2017;**224**:224-234. DOI: 10.1016/j.envpol.2017.01.083

[45] Halotech DNA. Leaflet of Oxisperm. Madrid, Spain: Halotech DNA; 2020. Ref: 04/2020 01

[46] BRED Life Science Technology Inc. Brochure: Seminal Reactive Oxygen Species (ROS) Detection Kits. Shenzhen, China: BRED Life Science Technology Inc; 2024

[47] Pahune PP, Choudhari AR, Muley PA. The total antioxidant power of semen and its correlation with the fertility potential of human male subjects. Journal of Clinical and Diagnostic Research. 2013;**7**(6):991-995. DOI: 10.7860/JCDR/2013/4974.3040

[48] Northwest Life Science Specialties. NWLSSTM Malondialdehyde Assay. Northwest Life Science Specialties; 2024. Ref: nwkmda01rev111507a

[49] G-Biosciences. Leaflet of MDA TBARs Assay. St. Louis, USA: G-Biosciences; 2024. Ref: #BAQ068

[50] Tavilani H, Doosti M, Saeidi H. Malondialdehyde levels in sperm and seminal plasma of asthenozoospermic and its relationship with semen parameters. Clinica Chimica Acta. 2005;**356**:199-203. DOI: 10.1016/j.cccn.2005.01.017

[51] Halotech DNA. Leaflet of Halosperm G2. Madrid, Spain: Halotech DNA; 2020. Ref: HT-HSG2

[52] Chohan K, Griffin J, Lafromboise M, De Jonge C, Carrell D. Comparison of chromatin assays for DNA fragmentation evaluation in human sperm. Journal of Andrology. 2006;**27**(1):53-59. DOI: 10.2164/jandrol.05068

[53] Evenson DP. The sperm chromatin structure assay (SCSA). And other sperm DNA fragmentation tests for evaluation of sperm nuclear DNA integrity as related to fertility. Animal Reproduction Science. 2016;**169**:56-75. DOI: 10.1016/j.anireprosci.2016.01.017

[54] Sharma R, Ahmad G, Esteves SC, Agarwal A. Terminal deoxynucleotidyl transferase dUTP nick end labeling (TUNEL). Assay using bench top flow cytometer for evaluation of sperm DNA frag-mentation in fertility laboratories: Protocol, reference values, and quality control. Journal of Assisted Reproduction and Genetics. 2016;**33**(2):291-300. DOI: 10.1007/s10815-015-0635-7

[55] Dimitriadis F, Borgmann H, Struck JP, Salem J, Kuru TH. Antioxidant supplementation on male fertility-A systematic review. Antioxidants. 2023;**12**(4):836. DOI: 10.3390/antiox12040836

[56] Neha K, Haider M, Pathak YM. Medicinal prospects of antioxidants: A review. European Journal of Medicinal Chemistry. 2019;**178**:687-704. DOI: 10.1016/j.ejmech.2019.06.010

Chapter 6

The Effect of Reactive Oxygen Species on Modulation of Structural and Functional Properties of Human Lymphocytes

Marina A. Nakvasina and Valery G. Artyukhov

Abstract

The effect of reactive oxygen species (ROS) on functional properties (the level of cytotoxic activity, antibody-forming ability, synthesis of tumor necrosis factor α, and free cytosolic calcium), the structural state of membranes, assessed based on the level of some surface markers, and apoptosis processes in lymphocyte cells in donor peripheral blood were studied. It was revealed that $O_2^{\bullet}$, $OH^{\bullet}$, H_2O_2, and $^1O_2$ have a predominantly inactivating effect on the level of cytotoxic activity of lymphocytes in relation to Ehrlich ascitic carcinoma cells, IgG synthesis, as well as the expression of receptors and surface markers: Fc receptors, CD3, CD19, CD56. It has been established that the level of intracellular calcium, which is a secondary messenger, increases in lymphocytes exposed to H_2O_2 (10^{-6} mol/l), $^1O_2$, and $OH^{\bullet}$, compared to intact cells. Compared to lymphocytes modified in a calcium-free medium, the presence of calcium ions in the lymphocyte suspension medium induces an increase in the number of cells at the early and late stages of apoptosis 6 hours after exposure to H_2O_2 and $^1O_2$. Apoptotic death of lymphocytes in a calcium-containing medium, induced by exposure to ROS involves the p53-dependent pathway of the nuclear mechanism, as well as receptor-mediated caspase and mitochondrial mechanisms of programmed cell death.

Keywords: reactive oxygen species, lymphocytes, structural and functional state, apoptosis, Ca^{2+}

1. Introduction

In the development of a specific immune response, the dominant role belongs to lymphocyte cells, able to selectively recognize their own and foreign antigens and then develop an active reaction aimed at neutralization of antigens. Lymphocytes are directly involved in the neoplastic process associated with the recognition of transformed cells, their neutralization and elimination from the body [1, 2]. One of the priority areas of biomedical research is the search and development of methods for targeted activation of the immune response to a developing tumor and enhancing the antigenic properties of neoplastically transformed cells.

At the moment, the role of reactive oxygen species (ROS) and activated oxygen metabolites (AOM) is actively being discussed in relation to the damage to tissues, organs, cells and molecules in the implementation of the functions of phagocytes and lymphocytes in the regulation of vascular tone, carcinogenesis, in inflammatory processes, ontogenesis and cellular proliferation, and aging [3–8]. Possible mechanisms of the tumor-promoting action of ROS are being investigated at the second stage of carcinogenesis – the promotion stage realized *via* the stimulation of proliferation, changes in intercellular interactions, stimulation of cell migration, and blocking apoptosis [9–11]. The inflammation is one of the factors increasing the risk of malignant neoplasms, and the alteration stage of inflammation is associated with intensive formation of ROS. Considering the complex nature of interactions between tumor cells and immunocytes of the organism with the tumor, significantly influenced by microenvironmental factors (such as ROS and calcium ions) [12, 13], model studies investigating the features of the functioning of immunocompetent and transformed cells under modulating conditions of these factors are necessary.

The implementation of the cytotoxic activity of lymphocytes in vitro is associated with T-lymphocyte-dependent antibody-independent and antibody-dependent mechanisms [1, 14, 15]. T-lymphocyte-dependent antibody-independent mechanism is provided by cytotoxic $CD8^+$ T-lymphocytes. Antibody-dependent cellular cytotoxicity is realized with the involvement of natural killer cells with FcγRIII receptor, which binds to the target cell – an immunoglobulin G antibody complex.

The term "Fc receptor" refers to a number of membrane-bound antigens (CD16, CD32, and CD46) interacting with a constant region of the immunoglobulin macromolecule [1]. Through Fc receptors, antibodies recruit leukocytes to degrade and eliminate the antigen recognized by the antibody. CD16 is a component of the low-affinity Fc receptor for IgG, FcγRIII, which stimulates antibody-dependent cellular cytotoxicity driven by NK. FcγRIII is associated in the cell cytoplasm with the ζ-chain of CD3, which conducts a signal inside the cell that activates the cytotoxic apparatus of natural killer cells. CD8 is a marker of a subpopulation of T-lymphocytes, cytotoxic lymphocytes, which is a coreceptor for T-cell receptor binding to MNS-1. $CD8^+$ lymphocytes destroy tumor and virus-infected cells, and participate in allograft rejection mechanisms, and autoimmunity reactions. CD19 is an additional signal-transducing molecule associated with the B-cell antigen-recognition receptor (ARR). CD56 is an exclusive marker of natural killer cells, can be expressed on the surface of mature T-lymphocytes, and is involved in adhesion and activation processes.

In order to reveal the contribution of each of the above-mentioned mechanisms of lymphocyte cytotoxicity, it seems necessary to investigate the activity of T- and B cells in relation to tumor cells under conditions of exogenous generation of reactive oxygen species, as well as to establish the relationship between changes in the functional properties of immune cells and the processes of modification of individual components of their membranes under conditions of ROS generation.

Elucidation of the possible effects of ROS and AOM on lymphocytes and their components are needed for the establishment and understanding of the patterns of functioning of immunocompetent cells and regulation of their properties and for the development of immunocorrection methods for complex therapy of socially significant human diseases, expanding modern ideas about purposeful control of programs of cell vital activity under pathology and exposure to extreme environmental factors. The aim of our research was to investigate the functional properties (the level of cytotoxic activity, antibody-forming ability, the synthesis of tumor necrosis factor α, and free cytosolic calcium), the structural state of membranes, assessed based on the level of some surface markers, and the apoptosis processes of lymphocytes in donor peripheral blood after exogenous generation of reactive oxygen species – superoxide anion radical, hydroxyl radical, singlet oxygen, and the addition of hydrogen peroxide.

2. Materials and methods

The object of the study was lymphocyte cells extracted from heparinized donor blood. The donors were healthy males aged 18–45. The blood samples were collected at the Blood Service of Voronezh Clinical Emergency Hospital No. 1.

Lymphocytes were obtained by centrifugation of heparinized donor blood using the density gradient of Ficoll-Urografin (7.64 g of ficoll 400 (Diam, Switzerland), 20 ml of 76% urografin (Bayer Schering Pharma AG, Germany), 92.56 ml of water; $\rho = 1.077\ g/cm^3$) [16]. In the centrifuge tubes, a 1:1 mixture of 3 ml of heparinized blood and Hank's Balanced Salt Solution (0.1 mol/l, pH 7.4 at 20°C) was layered over 1 ml of the gradient mixture. The mixture was centrifuged for 15 minutes at 200 g. The obtained lymphocytes were collected using a Pasteur pipette and were then centrifuged in excess of Hank's Balanced Salt Solution for 10 minutes at 180 g. Washed lymphocytes were suspended in Hank's Balanced Salt Solution until they reached a certain concentration using a hemocytometer.

To assess the purity of the obtained lymphocyte suspension, samples were examined by means of Romanowsky–Giemsa staining [16]. The proportion of lymphocytes in the obtained suspension was 95–97% of all cells, with 3–5% of cells being neutrophils and monocytes.

The separation of lymphocyte cells into T- and B-populations was carried out according to the Terasaki method [16]. The purity of T and B-cell suspensions obtained using the Terasaki method was estimated using direct immunofluorescence [16]. The composition of a cell suspension (mixture) of lymphocytes included 65% T cells, 17% B cells, and 18% natural killer cells; the composition of T-cell suspension included 87% T cells, 4% B cells, and 9% natural killer cells.

The following systems for generating reactive oxygen species were used in the study: hydroxyl radical ($OH^•$) was generated by the system of univalent reduction of hydrogen peroxide with metal ions – to 0.9 ml of lymphocyte suspension (2×10^6 cells/ml) was added 0.05 ml of aqueous solution of $FeSO_4$ at a concentration of 5×10^{-5} mol/l and incubated for 10 min at room temperature, then 0.05 0.02% solution of H_2O_2 was added [17]; superoxide anion radical ($O_2^•$) was produced in the tetramethylethylenediamine – riboflavin system under the visible light – 0.9 ml of lymphocyte suspension (2×10^6 kl/ml) was supplemented with 0.05 ml of riboflavin solutions (1.3×10^{-4} mol/l) and TEMED (0,1 mol/l prepared with 0.002 mol/l EDTA solution), simultaneously with the addition of TEMED, the system was irradiated with incandescent light (60 W power, 10 cm distance to the sample) for 1 min [17]; singlet oxygen ($^1O_2$) was generated in the presence of methylene blue (10^{-6} mol/l) under irradiation of 1 ml of lymphocyte suspension (2×10^6 cells/ml) with red light using the Ulox device, (Voronezh branch of the state research and production company Microtek), maximum radiation wavelength – 665 ± 15 nm, radiation range – 630–700 nm, output radiation intensity – 20 mW/cm^2. Hydrogen peroxide (H_2O_2) was used at a final concentration of 10^{-6} and 10^{-5} mol/l.

The suspension was incubated in an RPMI-1640 medium (BioloT, Russia) at 37°C in an atmosphere of 5% CO_2.

The cytotoxic activity of lymphocytes towards Ehrlich ascitic carcinoma cells cultured in laboratory mice (SHK line) was determined using colorimetric MTT analysis [18]. Ehrlich ascites carcinoma (EAC) was cultured in laboratory mice of SHK line. After injection into the abdominal cavity of the animal, the carcinoma developed within 7 days. The mouse was then killed, and ascites fluid was collected. The cells were then freed from EAC fluid by centrifugation for 5 min at 3000 rpm and washed in excess of Hanks' solution under the same conditions. The cells were suspended to a concentration of (2.5×10^4 cells/ml) in a sterile culture medium RPMI-1640 with L-glutamine.

The antibody-forming ability of human lymphocytes was assessed using local haemolysis [16]. The viability of lymphocytes was determined by standard methods using trypan blue [16].

The level of the expression of Fc receptors, CD3, CD8, CD19, and CD56 markers on the surface of native and modified lymphocytes was determined by means of the enzyme-linked immunosorbent assay (ELISA) using test systems (Sorbent, Moscow). Determination of the concentration of tumor necrosis factor (TNFα) in lysates of native and modified lymphocytes was carried out using an ELISA test system (Protein Contour Ltd., St. Petersburg).

To determine the number of apoptotic cells in the lymphocyte suspension by means of flow cytophotometry, a CyFlow Space flow cytometer (Partec, Germany) and an Annexin V-FITC Kit for the detection of apoptosis (Beckman Coulter, France).

DNA was isolated from lymphocytes using the phenol method. The fragmentation and the degree of DNA damage were analyzed by means of electrophoresis in agarose gel [19], and by means of the comet assay [20, 21]. Comet scanning was carried out using the CometScore™ software (http://www/autocomet.com/products_cometscore.php). The percentage composition of DNA in the comet trail (% T), characterizing the degree of the DNA damage in the studied cells, which can be considered as a signal for the initiation of apoptosis, was calculated [22].

The level of CD95 expression on the surface of native and modified lymphocytes was determined using the enzyme immunoassay method using monoclonal LT95 antibodies and their conjugate – goat anti-mouse IgG antibodies labeled with horseradish peroxidase (Sorbent, Russian Federation), and the FITC Mouse Anti-Human CD95 kit (BD Pharmigen™, USA) using a Guava Easy Cyte 8HT flow cytometer (Merck Millipore), data processing was carried out using In Cyte software.

The enzymatic activity of caspases-3, −8, −9 in native and modified lymphocytes was determined by the luminescent method [23] using a Shimadzu RF-1501 spectrofluorimeter (Japan) in accordance with the protocols of the manufacturers of the fluorescent substrates used in the study: for caspase-3 – N-acetyl-aspartyl-glutamine-valin-aspartyl-7-amino-4-methylcoumarin (BD Biosciences Pharmingen, USA), for caspase-8 – acetyl-isoleucyl-glutamyl-threonyl-aspartyl-7-amino-4-trifluoromethylcoumarin (Biovision, USA), for caspase-9 – acetyl-leucine-glutamic acid-histidine-aspartic acid-7-amino-4-trifluoromethylcoumarin (Abcam, USA).

The level of p53 in the cytosol of intact and modified lymphocytes was determined by means of the enzyme-linked immunosorbent assay using Bender MedSystems kits (Austria).

The concentration of free cytoplasmic calcium in the lymphocytes was measured using a Fura-2AM fluorescence probe on a Shimadzu RF-1501 spectrofluorophotometer (Japan) [24].

The level of reactive oxygen species (ROS) in intact and modified lymphocytes was studied by means of fluorescence technique using 2′,7′-dichlorofluorescein diacetate (DCFH-DA) on a Shimadzu RF-1501 spectrofluorophotometer (Japan) [25].

The mitochondrial membrane potential was recorded using a fluorescent probe – rhodamine 123 (Fluka, Germany) on a Shimadzu RF-1501 spectrofluorophotometer.

Statistical processing of the results was performed using Statgraphics spreadsheets. The reliability of the difference in the compared results was determined using the Student's t-test at a significance level of 95%.

3. Results

A statistically significant decrease in the level of cytotoxic activity towards Ehrlich ascitic carcinoma cells and antibody-forming ability (IgG) of lymphocytes, respectively, after exposure to $OH^\bullet$, $O_2^\bullet$, H_2O_2 (10^{-6} mol/l), and H_2O_2 and $OH^\bullet$(10^{-6} mol/l) (**Table 1**).

The superoxide anion radical and hydroxyl radical in lymphocytes stimulated the formation of the cytokine-tumor necrosis factor-alpha (TNFα), the initiator of apoptosis in tumor cells, while hydrogen peroxide inhibited the accumulation of TNFα (**Table 1**).

A decrease in the level of following membrane receptors, mediating reactions of cytotoxic activity, was revealed after exposure of lymphocytes to $O_2^\bullet$, OH, H_2O_2 compared with that of native cells (**Table 1**): Fc receptors (CD16, CD32, and CD46), CD3, CD19, and CD56. The level of expression of CD8, a coreceptor for the binding of the T-cell receptor to MHC-1, did not change under exposure to ROS.

Apparently, the studied ROSs directly or through products of lipid peroxidation and oxidative modification of proteins cause disruption of the structural and functional state of T- and B-lymphocyte receptors for antigen and other protein components of the external signal transduction cascade in the cell. The decrease in the level of expression of the tested membrane markers after oxidative modification of immunocytes is probably related to receptor shedding or their immersion in the modified membrane [26].

It was shown that $^1O_2$, $OH^\bullet$, and H_2O_2 induce an increase in the level of the universal second messenger-free calcium in lymphocytes suspended in a calcium-containing medium and in a "calcium-free" medium (Hank's solution, not containing Ca^{2+}), relative to that for native cells (**Table 1**). In the presence of verapamil (10^{-6} mol/l), blocker of calcium channels of plasma membranes in Ca^{2+}-containing medium, there was a statistically significant decrease in Ca^{2+} concentration in cells modified by the action of singlet oxygen and hydroxyl radical, respectively, up to 306.0 ± 64.0 nmol/l and 339.7 ± 34.1 nmol/l compared with that for lymphocytes after ROS generation in the absence of verapamil. Consequently, an increase in the level of cytosolic calcium in lymphocytes under the exposure to ROS was associated with the release of Ca^{2+} from intracellular depots, calcium entry from the medium via Ca^{2+} channels inhibited by verapamil, a violation of the structural state of plasma membranes due to the intensification of lipid peroxidation.

The level of free ionized calcium in the cytoplasm is an important component of the induction of apoptosis in many cell types [27]. Calcium ions modulate the processes of proliferation of immune and tumor cells, as well as the effector functions of immunocytes [13, 27].

Quantitative parameters characterizing the apoptotic death of lymphocytes after exposure to $^1O_2$ and H_2O_2 in the presence and absence of exogenous calcium in the cell suspension medium were determined using flow cytometry (**Figures 1–3**).

Parameter	Samples				
	Control	$O_2^{\bullet}$	H_2O_2	$OH^{\bullet}$	$OH^{\bullet}$
Cytotoxic activity of a mixture of lymphocytes, %	100	91.2 ± 2.0*	68.5 ± 0.6*	95.8 ± 1.6*	99.7 ± 0.8
Cytotoxic activity of T-cell suspension, %	100	90.5 ± 3.0*	85.5 ± 4.3*	88.7 ± 3.1*	93.6 ± 1.2*
Antibody-forming capacity of B-lymphocytes (by IgG), %	100	152.0 ± 8.9*	67.0 ± 15.6*	70.0 ± 13.3*	100.0 ± 20.0
Level of TNFα production, pg./ml	50.1 ± 0.8	60.1 ± 0.9*	44.2 ± 0.5*	60.7 ± 1.2*	50.3 ± 0.1
Level of expression of Fc receptors, ODU	0.50 ± 0.01	0.46 ± 0.01*	0.41 ± 0.01*	0.43 ± 0.01*	0.31 ± 0.02*
CD3 expression level, ODU	0.46 ± 0.01	0.36 ± 0.01*	0.34 ± 0.02*	0.38 ± 0.01*	0.42 ± 0.01*
CD8 expression level, ODU	0.38 ± 0.01	0.40 ± 0.01	0.40 ± 0.02	0.38 ± 0.01	0.39 ± 0.01
CD19 expression level, ODU	0.50 ± 0.01	0.39 ± 0.02*	0.38 ± 0.01*	0.47 ± 0.01*	0.47 ± 0.02
Expression level, ODU	0.30 ± 0.01	0.21 ± 0.01*	0.23 ± 0.02*	0.26 ± 0.01*	0.27 ± 0.02
Level of intracellular calcium of lymphocytes suspended in Ca^{2+}-containing medium	121.9 ± 15.7	—	260.0 ± 43.1*	436.3 ± 48.0*	521.2 ± 61.2*
Intracellular calcium level of lymphocytes suspended in a "calcium-free" medium	114.6 ± 32.7	—	241.2 ± 71.2*	226.2 ± 20.8*	275.0 ± 77.5*

* *differences from the control were statistically significant.*

Table 1.
Changes in parameters characterizing the structural and functional state of lymphocytes after exposure to reactive oxygen species.

The effect of H_2O_2 on lymphocytes suspended in Ca^{2+}-containing and "calcium-free" media with their subsequent incubation for 1 hour did not lead to significant changes in the number of cells at the early stage of apoptosis and the late stage of apoptosis and necrosis (**Figure 1**) compared to intact cells. The number of cells at the early stage of apoptosis (4.12%) slightly increased compared to the control (0.80%) in the Ca^{2+}-containing medium when $^1O_2$ was generated. After the addition of H_2O_2 to cells suspended in Ca^{2+}-containing and "calcium-free" media followed by incubation for 3 hours, there was no significant increase in the number of cells at the early stage of apoptosis and the late stage of apoptosis or necrosis (**Figure 2**) compared to unmodified cells. The number of cells modified by exposure to $^1O_2$ after 3 hours of incubation in a Ca^{2+}-containing medium at an early stage of apoptosis increased to 14.14% compared to the control (1.8%). The flow cytometric analysis of lymphocytes modified by exposure to $^1O_2$ and H_2O_2 revealed that after 6 hours of incubation in Ca^{2+}-containing medium, the number of cells at the early stage of apoptosis was 20.87 and 9.08%, respectively (**Figure 3**), while in the "calcium-free" medium it was 6.23 and 1.63% (in control after 6 hours in a calcium-containing medium this value was 7.90%, in a "calcium-free" medium it was 2.35%).

For the identification of the mechanisms and pathways for the implementation of apoptotic death of lymphocytes under exposure to ROS, other marker indicators of

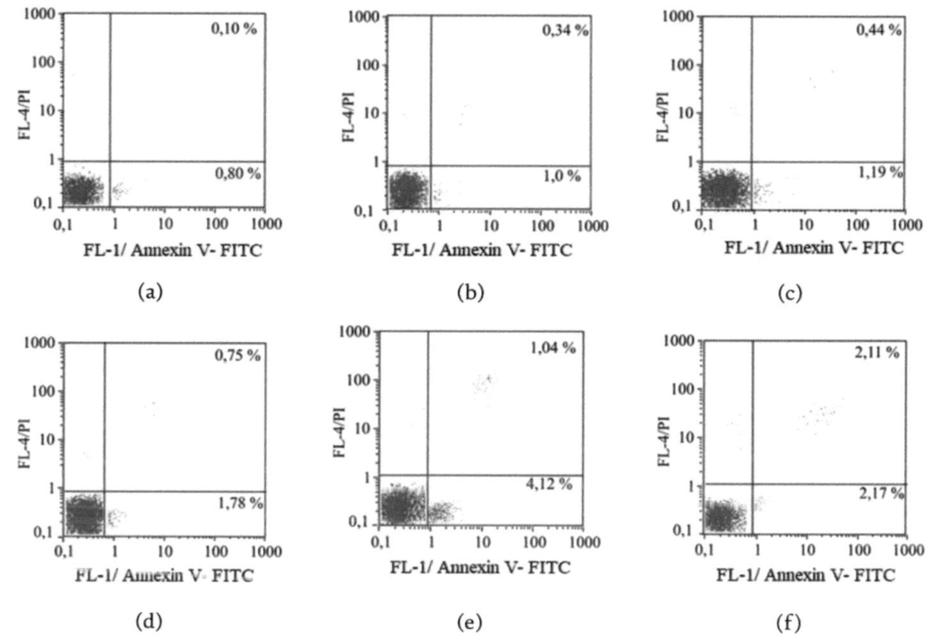

Figure 1.
Flow cytometric analysis of lymphocytes 1 hour after modification: a. control in Ca^{2+}-containing medium; b. control in a calcium-free medium; c. H_2O_2 in a calcium-containing medium; d. H_2O_2 in a calcium-free medium; e. $^1O_2$ in a calcium-containing medium; f. $^1O_2$ in a calcium-free medium.

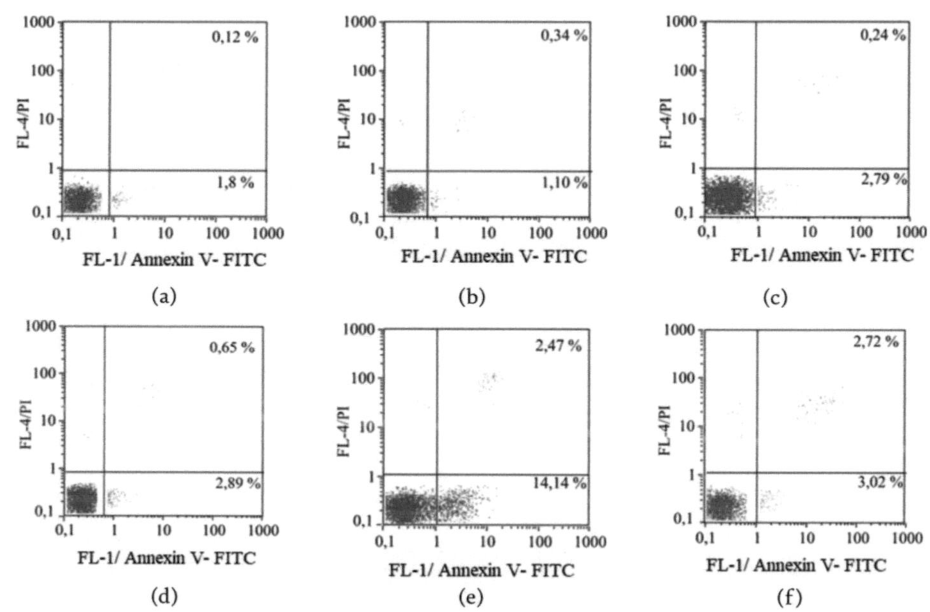

Figure 2.
Flow cytometric analysis of lymphocytes 3 hours after modification: a. control in Ca^{2+}-containing medium; b. control in a calcium-free medium; c. H_2O_2 in a calcium-containing medium; d. H_2O_2 in a calcium-free medium; e. $^1O_2$ in a calcium-containing medium; f. $^1O_2$ in a calcium-free medium.

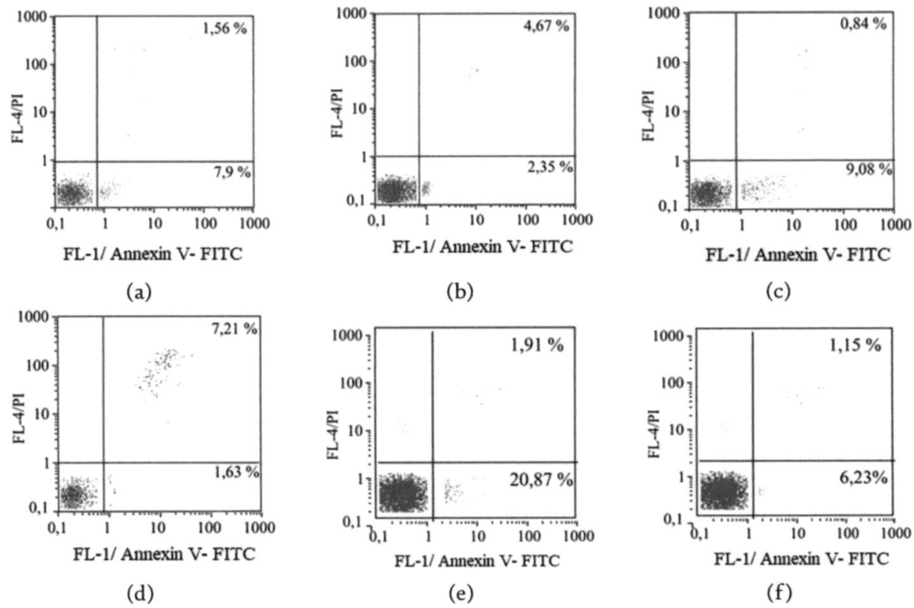

Figure 3.
Flow cytometric analysis of lymphocytes 6 hours after modification: a. control in Ca^{2+}-containing medium; b. control in a calcium-free medium; c. H_2O_2 in a calcium-containing medium; d. H_2O_2 in a calcium-free medium; e. $^1O_2$ in a calcium-containing medium; f. $^1O_2$ in a calcium-free medium.

this type of programmed cell death were dynamically studied: DNA fragmentation processes and the level of its damage, changes in the level of the transcription factor p53, death receptors Fas (CD95), caspases-8, −9, −3, reactive oxygen species, and calcium ions.

It was found that 20 hours after the exposure of lymphocytes to $O^{\bullet}_2$, $^1O_2$, $OH^{\bullet}$, H_2O_2 (10^{-6} mol/l), the sizes of DNA fragments varied, respectively, within the range of 2500–5000 bp, 1500–2500 bp, 2000–4500 bp, and 1500–4000 bp [28, 29]. The formation of large DNA fragments indicated the possibility of initiation of apoptosis pathways with the involvement of the transcription factor p53, the proapoptotic factor Bax, and the mitochondrial factor AIF. In the investigation of changes in the structural state of DNA, a significant increase of DNA damage (single-strand breaks) was recorded in comparison with the control immediately and 6 hours later (the maximal damage) after exposure of the cell suspension to hydrogen peroxide (10^{-6} mol/l). The addition of hydrogen peroxide (10^{-6} mol/l) caused a statistically significant increase in p53 concentration compared to intact cells [28].

During 1–5 hours after exposure of lymphocytes to superoxide anion radical, hydroxyl radical, singlet oxygen, and hydrogen peroxide (10^{-6} mol/l), a statistically significant increase in the expression level of membrane death receptors (CD95) was revealed compared to intact cells [28].

Using the flow cytometric method, it was found that hydrogen peroxide at a final concentration of 10^{-5} mol/l after 1 hour caused an increase in the expression level of CD95 lymphocytes by 17% and after 2 hours by 76% compared to control samples.

A 25% increase in the level of activity of initiating caspase-8 compared to that for unmodified lymphocytes was revealed immediately after cell modification with H_2O_2 (10^{-6} mol/l). An increase in the activity of effector caspase-3 compared to the control was found after 2 and 4 hours of incubation of lymphocytes modified by exposure to H_2O_2 in a concentration of 10^{-6} mol/l [28].

A statistically significant increase in the level of intracellular reactive oxygen species was revealed 1 and 2 hours after modification of cells with H_2O_2 (10^{-5} mol/l) compared to that for intact samples. A significant increase (2.6 times) compared to the control (intact cells) in the activity of initiating caspase-9, involved in the initiation of the mitochondrial caspase pathway of apoptosis, was detected 1 and 2 hours after exposure to H_2O_2(10^{-5} mol/l) [28].Caspase-9 further activated the effector caspase-3, which cleaves target proteins.

The exposure of immunocytes to hydrogen peroxide (10^{-5} mol/l) first induced an increase in the level of cytosolic calcium to 150 ± 18 nmol/l relative to that for intact cells (81 ± 15 nmol/l) and after 1 and 3 hours, a decrease, to 45 ± 8 and 34 ± 10 nmol/l respectively, compared to native samples (67 ± 10 and 59 ± 8 nmol/l) was found. It is possible that under these conditions, the redistribution of calcium ions between cellular compartments (cytosol, mitochondria, and nucleus) occurs in cells. The uptake of cytosolic calcium by mitochondria promotes the initiation of the mitochondrial mechanism of apoptosis. Ca^{2+} is required for the activation of Ca^{2+}, Mg^{2+}-dependent endonuclease catalyzing DNA fragmentation processes.

A 32% decrease in the fluorescence intensity of the "mitochondrial" fluorescent probe – rhodamine 123 was found compared to that for native cells (100%) 1 hour after modification with hydrogen peroxide (10^{-5} mol/l), indicating the impairment of the functional state of mitochondria.

The obtained results support the idea of the implementation of the p53-dependent pathway of nuclear, receptor-mediated caspase, and mitochondrial mechanisms of lymphocyte apoptosis under exposure to reactive oxygen species.

4. Conclusions

Summarizing our results, we can conclude that O_2, $O_2^{\bullet}$, $OH^{\bullet}$, and H_2O_2 have a predominantly inactivating effect on the majority of the studied indicators of the state of individual components of lymphocyte cells responsible for antitumour protection. A decrease in the cytotoxic activity of lymphocytes towards Ehrlich ascitic carcinoma cells, the level of synthesis of immunoglobulin G, and the expression of receptors and surface markers: Fc receptors, CD3, CD19, and CD56 were recorded in comparison with normal values. The processes of oxidative modification of lymphocyte membrane components led to changes in the level of expression of receptors, secondary messengers and, ultimately, the functional properties and death of the studied cells. A significant increase in the level of intracellular free calcium compared to the normal value, depending on the level of exogenous calcium, observed after exposure of lymphocytes to ROS triggered the changes in the functional properties of lymphocytes and their apoptotic death. The presence of calcium in the lymphocyte suspension medium under conditions of exogenous ROS generation accelerated the processes of apoptotic cell death. This can be explained by the presence of a calcium-sensitive receptor (CaSR) bound with G protein in the lymphocyte membranes, which plays an important role in the processes of cell differentiation, proliferation, apoptosis, and cytokines secretion [30].

Stimulation of the formation of reactive oxygen metabolites in cells under conditions of exposure to inducing reactive oxygen species is called "the phenomenon of ROS-induced ROS formation" [31, 32]. Small amounts of inducing ROSs cause a drop in the transmembrane potential of mitochondria and generation of secondary reactive oxygen species, which leads to the development of an "oxidative burst". In the course of ROS-induced ROS formation, proteins and regulatory thiols are oxidized, which changes the redox status of cells, initiating non-specific permeability of mitochondrial membranes: the functioning of the electron-transport chain is disturbed, and the properties of the mitochondrial membrane are altered.

An increase in the level of calcium in the cell promotes the mitochondrial apoptosis cascade, accompanied by the formation of transit pores and the release of cytochrome c and endonuclease G into the cytosol. The release of cytochrome c leads to a dramatic increase in intracellular AFC content and activation of the caspase cascade. When calcium levels increase in the cytosol of cells, apoptosis can be triggered with the participation of caspase-12, which activates effector caspases.

ROS can participate in triggering apoptosis not only by initiation of mitochondrial mechanism but also by increasing intracellular calcium concentration, formation of cAMP, DNA oxidation with subsequent activation of p53, oxidation of plasma membrane phosphatidylserine and inactivation of scramblase, increased protein phosphorylation leading to changes in the regulation of gene expression.

Considering the fact that at certain stages of tumor development, there is a significant intensification of free radical processes [8–11], we can conclude that reactive oxygen species, along with problastoma factors that suppress immunity and enhance tumor growth, will inhibit the functions of lymphocyte cells at the site of pathology. Probably, in this case, protection of immunocytes from oxidative stress is required.

We revealed that processes of apoptotic death were initiated in lymphocytes under conditions of generation of reactive oxygen species ($O^{\bullet}_2$, $^1O_2$, and $OH^{\bullet}$) and after the addition of exogenous hydrogen peroxide (at final concentration 10^{-6} and 10^{-5} mol/l), which were implemented with the involvement of receptor-mediated caspase, mitochondrial (with activation of caspase-9) and p53-dependent pathways of the nuclear mechanisms of apoptosis. It is possible that systems of the generation of reactive oxygen species and pathways for the redistribution of free calcium ions between cellular compartments can act as "sensitive centres" for the regulation of H_2O_2-induced apoptosis.

The identified relative resistance of lymphocytes to the action of singlet oxygen in relation to the studied indicators of the structural and functional state of cells (cytotoxic activity of lymphocytes, their antibody-forming ability, TNFα level, and expression of CD3, CD19, and CD56) can be an important fact when analyzing the mechanisms of photodynamic therapy of tumors, since it will ensure stable functioning of immunocytes under conditions of $^1O_2$ generation at the site of pathology.

Our results expand modern understanding of changes in parameters (indicators of the structural and functional state) of human peripheral blood lymphocytes, characterizing their participation in the implementation of antitumour immunity under exposure to reactive oxygen species. These results must be taken into account when discussing issues related to the development of methods for correcting the state of human immunocompetent cells during malignant cell growth and pathologies associated with the intensification of free radical processes.

Author details

Marina A. Nakvasina* and Valery G. Artyukhov
Voronezh State University, Russia

*Address all correspondence to: nakvasina_ma@mail.ru

References

[1] Khaitov RM, Ignatieva GA, Sidorovich IG. Immunology. Moscow: Medicine; 2000. 432 p

[2] Finn OJ. Immuno-oncology: Understanding the function and dysfunction of the immune system in cancer. Annals of Oncology. 2012;**23**(Suppl. 8):viii6-viii9. DOI: 10.1093/annonc/mds256

[3] Weidinger A, Kozlov AV. Biological activities of reactive oxygen and nitrogen species: Oxidative stress versus signal transduction. Biomolecules. 2015;**5**:472-484. DOI: 10.3390/biom5020472

[4] Alfadda AA, Sallam RM. Reactive oxygen species in health and disease. Journal of Biomedicine and Biotechnology. 2012;**1**:936486. DOI: 10.1155/2012/936486

[5] Winterbourn CC, Kettle AJ, Hampton MB. Reactive oxygen species and neutrophil function. Annual Review of Biochemistry. 2016;**85**:765-792. DOI: 10.1146/annurev-biochem-060815-014442

[6] Sies H, Jones DP. Reactive oxygen species (ROS) as pleiotropic physiological signaling agents. Nature Reviews Molecular Cell Biology. 2020;**21**(7):363-383. DOI: 10.1038/s41580-020-0230-3

[7] Redza-Dutordoir M, Averill-Bates DA. Activation of apoptosis signaling pathways by reactive oxygen species. Biochimia et biophysica acta. 2016;**1863**(12):2977-2992. DOI: 10.1016/j.bbamcr.2016.09.012

[8] Tafani M, Sansone L, Limane F, Arcangeli T, De Santis E, Polese M, et al. The interplay of reactive oxygen species, hypoxia, inflammation, and sirtuins in cancer initiation and progression. Oxidative Medicine and Cellular Longevity. 2016;**1**:3907147. DOI: 10.1155/2016/3907147

[9] Kobliakov VA. Mechanisms of tumor-promoting action of reactive oxygen species. Biochemistry (Moscow). 2010;**75**(6):757-769

[10] Galadari S, Rahman A, PallichankandyS, ThayyullathilF. Reactive oxygen species and cancer paradox: To promote or to suppress? Free Radical Biology and Medicine. 2017;**104**:144-164. DOI: 10.1016/j.freeradbiomed.2017.01.004

[11] Aggarwal V, Tuli HS, Varol A, Thakral F, Yerer MB, Sak K, et al. Role of reactive oxygen species in cell in cancer progression: Molecular mechanisms and recent advancements. Biomolecules. 2019;**9**(11):735. DOI: 10.3390/biom9110735

[12] Chen X, Song M, Zhang B, Zhang Y. Reactive oxygen species regulate T cell immune response in the tumor microenvironment. Oxidative Medicine and Cellular Longevity. 2016;**1**:1580967. DOI: 10.1155/2016/1580967

[13] Chen Y-F, Chen Y-T, Chiu W-T, Shem M-R. Remodeling of calcium signaling in tumor progression. Journal of Biomedical Science. 2013;**20**(1):23. DOI: 10.1186/1423-0127-20-23

[14] Martinez-Lostao l., Anel A., Pardo J. How do cytotoxic lymphocytes kill cancer cells? Clinical Cancer Research. 2015;**21**(22):5047-5056. DOI: 10.1158/1078-0432.CCR-15-0685

[15] Seidel UJE, Schlegel P, Lang P. Natural killer cell mediated

antibody-dependent cellular cytotoxicity in tumor immunotherapy with therapeutic antibodies. Frontiers in Immunology. 2013;**4**:76. DOI: 10.3389/fimm.2013.00076

[16] Lymphocytes KJ. Methods. Moscow: Mir; 1990. p. 400

[17] Armstrong D, editor. Oxidative Stress Biomarkers and Antioxidant Protocols. Totowa, New Jersey: Humana Press; 2002. 320 p. DOI: 10.138/1592591736

[18] Stoev S, Denev S, Dutton M, Nkosi B. Cytotoxic effect of some mycotoxins and their combinations on human peripheral blood mononuclear cells as measured by the MTT assay. The Open Toxicology Journal. 2009;**2**:1-8. DOI: 10.2174/1875414700902010001

[19] Herrmann M, Lorenz HM, Voll R, Grunke M, Woith W, Kalden JR. A rapid and simple method for the isolation of apoptotic DNA fragments. Nucleic Acids Research. 1994;**22**(24):5506-5507. DOI: 10.1093/nar/22.24.5506

[20] Singh NP, Mc Coy MT, Tice RR, Schneider ELA. Simple technique for quantitation of low Lewels of DNA damage in individual cells. Experimental Cell Research. 1988;**175**:184-192. DOI: 10.1016/0014-4827(88)-90265-0

[21] Olive PL, Banath JP, Durand RE. Heterogenety in radiation-induced DNA damage and repair in tumor and normal cells measured using the "comet" assay. Radiation Research. 1990;**122**:86-94. DOI: 10.1667/rrav04.1

[22] Hartmann A, Agurell E, Beevers C, Brendler-Schwaab S, Burlinson B, Clay P, et al. Recommendations for conducting the *in vivo* alkaline comet assay. Mutagenesis. 2003;**18**:45-51. DOI: 10.1093/mutage/18.1.45

[23] Kaufmann S, Lee S-H, Meng X, Loegering D, Kottke T, Heinzing AJ, et al. Apoptosis-associated caspase activation assays. Methods. 2008;**44**:262-272. DOI: 10.1016/j.ymeth.2007.11.005

[24] Hirst RA, Harrison C, Hirota K, Lambert DG. Measurement of [Ca^{2+}] i in whole cell suspensions using fura-2. Methods in Molecular Biology. 2005;**312**:37-45. DOI: 10.1385/1-59259-949-4:037

[25] Rastogi RP, Singh SP, Häder DP, Sinha RP. Detection of reactive oxygen species (ROS) by the oxidant-sensing probe 2′,7′-dichlorodihydrofluorescein diacetate in the cyanobacterium Anabaena variabilis PCC 7937. Biochemical and Biophysical Research Communications. 2010;**397**(3):603-607. DOI: 10.1016/j.bbrc.2010.06.006

[26] Albanese J, Dainiak N. Modulation of intercellular communication mediated at the cell surface and on extracellular, plasma membrane-derived vesicles by ionizing radiation. Experimental Hematology. 2003;**31**(6):455-464. DOI: 10.1016/S0301-472x(03)00050-x

[27] Marchi S, Patergnani S, Missiroli S, Morciano G, Rimessi A, Wieckowski MR, et al. Mitochondrial and endoplasmic reticulum calcium homeostasis and cell death. Cell Calcium. 2018;**69**:62-72. DOI: 10.1016/j.ceca.2017.05.003

[28] Nakvasina MA, Holyavka MG, Artyukhov VG, Radchenko MS, LidokhovaOV.MechanismsofUV-induced human lymphocytes apoptosis. Biophysical Reviews. 2023;**15**(5):1257-1267. DOI: 10.1007/s12551-023-01142-w

[29] Artyukhov VG, Trubicina MS, Nakvasina MA, Solovieva EV. DNA fragmentation of human lymphocytes in dynamics of development of apoptosis induced by action of UV radiation and

reactive oxygen species. Cell and Tissue Biology. 2011;**5**(2):61-67. DOI: 10.1134/S1990519X11020039

[30] Wu C, Wu Q, Du J, Zeng J, Li T, Xu C, et al. Calcium-sensing receptor in the T lymphocyte enhanced the apoptosis and cytokine secretion in sepsis. Molecular Immunology. 2015;**62**(2):337-342. DOI: 10.1016/j.molimm.2014.08.007

[31] Zandalinas SI, Mittler R. ROS-induced ROS release in plant and animal cells. Free Radical Biology and Medicine. 2018;**122**:21-27. DOI: 10.1016/j.freeradbiomed.2017.11.028

[32] Zorov DB, Juhaszova M, Sollott SJ. Mitochondrial ROS-induced ROS release: An update and review. Biochimia et biophysica acta. 2006;**1757**(5-6):509-517. DOI: 10.1016/j.bbabior.2006.04.029

Chapter 7

Oxidative Stress in Neurodegenerative Diseases

Yasemin Gündüztepe

Abstract

The main focus of this extensive review is how the presence of too much oxidative stress and free radicals (ROS) leads to the appearance of multiple disorders in the nervous system as Parkinson's, Alzheimer's, Huntington's, and, multiple sclerosis. Contrary to the prediction of antioxidants related to *in vitro* and preclinical *in vivo* studies exhibiting neuroprotective effects, clinical trials in the treatment of neurodegenerative diseases have shown little evidence of such significant benefits. The reason for this disparity could lie in several factors; for instance, the optimal dosages, administration timing, and effectiveness of delivery methods vary from preclinical to clinical stage. One of the main considerations is the ability of agents to be delivered through the blood-brain barrier (BBB) and the study of advanced systems of nonverbal delivery, such as nanoparticles and Lipsomal formulations. The article adds the point that targeting endogenous defense, like creating the Nrf2 pathway and inhibiting ROS production through NOX enzyme inhibitors, can be promising strategies in alleviating oxidative stress concerning the CNS. The review calls for looking into alternatives to current therapeutic approaches, with emphasis on advancing two aspects, improving the delivery of the medicines and increasing the body's antioxidants, as key areas of future investigation.

Keywords: oxidative stress, neurodegenerative diseases, free radical, antioxidant, neuroinflammation

1. Introduction

Neurodegenerative diseases are disorders that share common features of neuronal dysfunction and death in a selective pattern, which results in extensive damage to the cerebral cortex and basal ganglia, with consequent severe deficits in memory and movement: Alzheimer's disease, Parkinson's disease, Huntington's disease, and multiple sclerosis. Such disorders significantly reduced physical health and led to disability, as well as the patient's death. These diseases are generated as a result of the interdependencies of many factors, which include oxidative stress at its core. This stress is also accompanied by a shift in the cellular redox state within the nervous system. As an oxidative stress, it refers to the condition whereby bodies generate reactive oxygen species (ROS) in ratios that are beyond the ability of defense mechanisms to counteract them. Some of the ROS are the superoxide anion radical (O_2^-), the hydroxyl radical ($OH^\cdot$), and the hydrogen peroxide (H_2O_2), which are all oxygen-containing

IntechOpen

molecules. ROS forms naturally in all living cells and is a direct byproduct of oxygen consumption, especially during the process of aerobic respiration in mitochondria.

Although low levels of ROS are involved in crucial cellular signaling mechanisms that maintain cellular homeostasis, significant amounts should be damaging to the cell structures such as lipids, proteins, and DNA. The brain for instance, which accounts for only about 2% of total body mass, is known to draw approximately 20% total body oxygen, due to its high metabolic rate that is highly susceptible to oxidative damage. Lipids are primarily a component of neuronal membranes since neurons are principally found in the brain, and they comprise highly peroxidable lipids. This process produces toxic lipid peroxides that when not broken circulate the oxidative cycle leading to the damage of neurons. It is found that aging results in the decline of antioxidant status and further deterioration of the mitochondrial membrane leading to higher production of ROS. Recent findings have supported the opinion that age-dependent increase in OS is a significant factor in the molecular mechanisms of neurodegenerative diseases. For example, oxidative stress is associated with amyloid plaque in AD, Lewy bodies in PD, and misfolded proteins in HD. This accords with the pathological features of synapse dysfunction, neuron degeneration, and, hence, cognitive impairments. Nevertheless, the acute processes where oxidative stress executes a deleterious effect on neurodegenerative diseases' development are not fully elucidated. They conclude that it is currently unknown and conflicted, especially when it comes to the precise mechanisms and the interplay between genes and the environment. Therefore, the goal of this paper is to give a comprehensive review of the involvement of oxidative stress in neurodegenerative diseases and conceptually discuss existing shortcomings and novel prospects for oxidative stress management as a potential approach to treating progressive neuronal damage in the context of various neurodegenerative diseases.

Alzheimer's disease, Parkinson's disease, Huntington's disease and other diseases of the neurological systems are among the cardinal diseases that have a profound impact on the elderly people. Although much effort has been devoted to the molecular dissection of these diseases, there is still a lot that is not known about how these diseases work, which is why there is a need to look for new ways to decipher these diseases. It is crucial to appreciate the intention of this study as providing a unique approach to the investigation of oxidative stress contribution to neurodegenerative diseases; thus, the findings of this research contribute unique knowledge enhancing the understanding of neurodegenerative diseases from a novel angle. While previous works have primarily emphasized the singular features of oxidative damage, thus enhancing our knowledge of different components of oxidative stress but not necessarily the links among them and its impact on neuronal degeneration, our work is novel in its attempt to employ various methodology straddling several fields to understand the diverse effects that oxidative stress has in neuronal degeneration. So, using modern molecular approaches and techniques in imaging, biochemistry, and computation, we have pinpointed new biomarkers and signaling pathways that de–/construct the relationship between oxidative stress and neurodegeneration. It becomes evident that oxidative stress is a powerful driver for the onset of neurodegenerative diseases, and our work has also shown that there are molecular pathways therein that have not been described before. This new concept has innovative implications that disrupt conventional framings and create new opportunities in the development of anti-therapeutic approaches for therapeutic solutions to reducing oxidative stress. In addition, our findings do not stop at the discovery of new mechanisms, as there seems to be transitions between higher and lower number states that have not

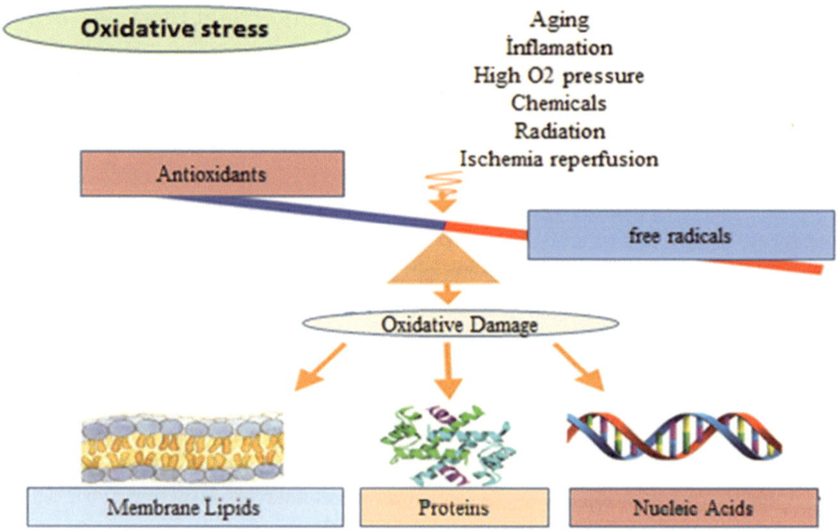

Figure 1.
Oxidative balance.

been previously described. By combining our new insights with the existing knowledge, we introduce a clear and coherent CCM framework that will be beneficial to the future development in the field. In addition to expanding the wealth of present-day knowledge, this approach creates the prerequisites for creating new diagnostic and therapeutic methods (**Figure 1**).

2. Literature reviews: development

The content of this study emphasizes that oxidative stress is a common pathological feature in neurological diseases, and ROS are the main mechanism of this cellular redox imbalance. Because antioxidants protect biological targets against ROS, they are being evaluated as potential therapeutic agents to prevent neuronal damage. However, although the *in vitro* and preclinical *in vivo* data are encouraging, the clinical effectiveness of antioxidant treatment strategies is limited, and most clinical trials using antioxidants as a therapeutic agent in neurodegenerative diseases have produced disappointing results. This may be due to the need to adjust the concentration and time parameters between declinical studies and clinical settings [1].

Moreover, considering that a successful therapeutic agent should easily cross the blood–brain barrier (BBB) for neurological diseases, new and effective delivery methods need to be investigated. In this context, the use of compounds that cross the BBB and strengthen the endogenous antioxidant defense mechanism, for example, by activating the Nrf2 pathway, or compounds that can modulate the production of ROS, such as NOX enzyme inhibitors, is considered a more promising approach to combating oxidative stress in the central nervous system (CNS). In this article, the main players of oxidative stress and the evidence for their role in Parkinson's disease, Alzheimer's disease, Huntington's disease, and multiple sclerosis are briefly summarized, and the therapeutic potential of antioxidants for central nervous system disorders is critically discussed, especially with an emphasis on new and innovative therapeutic strategies. More in-depth analyses can be carried out on clinical studies and preclinical models on why the effectiveness of antioxidant therapies on neurological diseases remains limited. In these studies, focusing on the optimization of parameters such as the dosage of antioxidants, the timing of administration, and the mode of delivery can provide important information for future studies.

Innovative drugs such as nanoparticle-dependent systems carrying transporters or liposomal formulations can direct drug delivery for the antioxidants to the brain. Furthermore, the investigation of molecules that can restructure either oxidative stress pathways or even ROS production may lead the field of research in this area. Activation of the Nrf2 pathway could safeguard against ROS at the site by way of upgrading the cellular antioxidant system. Therefore, Nrf2 activators that provide a supportive environment for the neurons may hold the key to the establishment of new treatment options. Moreover, suppression of the NOX enzymes by its inhibitor can also slow the course of diseases appearing within the framework of neurodegeneration.

This special issue focuses on the potential role of oxidative stress in neurodegenerative diseases and includes seven original research articles and two reviews that show how oxidative stress can contribute to neurodegeneration in various neuronal and disease contexts in various clinical and preclinical settings. Increases in ROS and RNS levels and impaired antioxidant defense mechanisms are often observed in most neurodegenerative diseases, such as Alzheimer's, Parkinson's, Huntington's disease, and amyotrophic lateral sclerosis (ALS) [2].

- Liou C. et al.: Memantine, a cholinesterase inhibitor therapy used in Alzheimer's disease (AD) patients, reduces the levels of oxidative thiobarbituric acid-related substances (TBARS) in blood samples of AD patients compared to non-AD controls. This study investigates whether individuals carrying the epsilon 4 allele of the APOE4 gene show high TBARS and low levels of the antioxidant thiol and whether it is associated with lower mitochondrial DNA copy counts.

- Nguyen C.D. and Lee G.: Melittin has demonstrated its neuroprotective mechanism in mouse hippocampal cell line and *in vivo* mouse model based on A Dec25–35 ICV injection. Melittin treatment regulates the nuclear translocation of Nrf2 and the production of the antioxidant enzyme heme oxygenase-1 (HO^{-1}), as well as improves the learning and memory performance of mice treated with Aß25–35.

- Wang W. et al.: This study describes the cross decoupling between glutamate and dopamine, which may regulate dopamine oxidation, which has a toxic effect on Parkinson's disease (PD). Among the mechanisms by which glutamate can inhibit dopamine oxidation are the inhibition of the auto-oxidation of dopamine and the copper-mediated oxidation of decathelamines.

- Mustafa R. et al.: This study focused on understanding the role of primary cilia in the protection of dopaminergic neurons under conditions of mitochondrial and oxidative stress-induced and basal conditions in a pharmacological mouse model of PD. By producing a line of mutant mice lacking primordial cilia, this study showed that the primary cilia can play an auto-inhibitory role in the spontaneous activity of dopaminergic neurons and that dopaminergic neurons are partially protected from the neurotoxic effect generated by the PD-toxin 1-methyl-4-phenyl-1,2,3,6-tetrahydropyridine (MPTP).

- Bobadilla M., Hernandez C., et al.: This study investigated the effect of oral administration of grape juice fortified with red grape polyphenols on the brains of mice under stress. Molecular readings, such as the increased expression of genes associated with combating oxidative stress, have shown the benefits of this newly developed drink.

The studies in this special issue offer new candidates for developing new models, potential treatment targets, and reliable biomarkers to understand the effects of oxidative stress on neurodegeneration. Despite the main antioxidant therapies that have failed in the past, these emerging new pathways and mechanisms can help us better understand the mechanisms of disease, and this field is actively developing and progressing towards identifying clinical candidates. It is hoped that this research collection will contribute valuable information to the race to eliminate neurodegenerative diseases. The article titled "Oxidative Stress and Antioxidants in Neurodegenerative Disorders" provides an overview of oxidative stress and the role it plays in the development of various neurodegenerative diseases and how antioxidants might, the rapeutically, bring the situation under control. This article discloses a mechanism through which oxidative stress is recognized as a principal culprit in the genesis and progression of neurodegenerative disorders like Alzheimer's disease, Parkinson's disease, ALS, and Huntington's disease. The article also explains that excessive oxidative stress is modifiable and hence can be reversed or managed [3].

Such an imbalance is characterized by the intracellular molecular processes involving the perishing of neurons and their death, which trigger the neurological symptoms of dementia. The writing piece investigates the oxidative stress role in some diseases like Alzheimer's disease, Parkinson's disease, ALS, and Huntington's disease among others. It is pointed out that the oxidative stress that comes with cell dysfunction is of great importance in these ailments. With many antioxidant-based therapeutic ones being discussed, there would be options to treat or even reverse oxidative stress. Through these novel treatments, new therapeutic methods with the ultimate purpose of preventing brain dysfunctional disorder and gaining confidentiality in the health status can be provided. Through a systematic review of the literature, this article also provides the latest data on the oxidative stress mechanisms that relate to the loss of pathophysiology of the neurodegenerative disease. The study shows in enhanced that oxidative stress plays a central role in the neurodegeneration process, and thus, supporting can suppress it, which may slow the progression of the disease. It is written that there is a need to focus on the research that could be done about the benefits and the accessibility of antioxidant therapies.

In the article "Oxidative stress in neurodegenerative disease," C. X. from the Department of Laryngology at Tongji Hospital and also K. J. Occupational Medicine at Zhongshan Hospital examine the role of oxidative stress in the pathophysiology of neurodegenerative diseases and the therapeutic strategies involved in this process. Of particular interest, the chapter on oxidative stress and its role in the process of neurodegeneration is described in detail including the production of (ROS among the cell population as well as the alteration of their prooxidant and antitoxic equilibrium. This article describes types of reactive oxygen species produced as a by-product during aerobic metabolism and are formed by many enzyme reactions with mitochondrial fatty acid oxidation. While internal biological processes activate the release of ROS, external factors of stress also can be the cause, such as ionizing radiation and oxidizing chemicals [4].

Briefly, they highlight that low concentration levels of ROS regulate intracellular signaling as a second messenger but if the exposure is prolonged or the concentration of ROS elevated, at that stage cell macromolecules such as proteins, lipids, and DNA are damaged and hence, both apoptotic and necrotic cell death are induced. It is pointed out that the neurodegenerative diseases' creation is followed by progressive cell loss mainly in certain groups of neurons, and the disorder of protein aggregation and metal ions equilibrium in this process is very closely related to the processes of oxidative stress (**Table 1**).

Multiple therapeutic techniques that are tailored to disorders with oxidative stress are described, which are in turn associated with different neurodegenerative

Title	Description and Content	Key Findings
Oxidative Stress and Neurodegenerative Diseases	This article also addresses the application of oxidative stress in diseases such as Parkinson's, Alzheimer's, Huntington's, and multiple sclerosis. Oxidative stress occurs because muscles cell have amassed active oxygen species (ROS) in their system.	It was noted that while connection between ROS and various diseases was established *in vitro* and in preclinical models, clinical trials failed to demonstrate an appropriated level of antioxidant supplementation effectiveness. Possible factors include antioxidant doses, its administration time, and whether it is able to penetrate the Blood-Brain Barrier (BBB).
Oxidative Balance and Reactive Oxygen Species (ROS)	Oxidative homeostasis is governed by the ability of the system to produce ROS and, at the same time, possess the ability to counteract it through the antioxidant system. Specifically, the brain is more at risk for ROS accumulation since it is well-known to metabolize a high level of oxygen.	The human brain consumes as much as 20% of the oxygen supplied to the human body and, thus, contributes to ROS generation. Increased levels of ROS can cause lipid peroxidation and harm to protein and DNA structures, which can ultimately cause cell death.
Nrf2 Pathway and NOX Enzyme Inhibitors	Potential the rapeutical targets, namely, the Nrf2 pathway, and NOX enzyme inhibitors may help to mitigate oxidative stress levels.	Nuclear factor erythroid-derived 2, Nrf2, when activated, assists in increasing the antioxidant capacity within cells. Abstract: This study provides a comprehension and understanding of how NOX enzyme inhibitors can help decrease ROS production and repair neurodegenerative diseases.
Clinical Studies and Antioxidant Therapies	These are just some of the inconsistencies that occurred in clinical trials involving antioxidants, which can be viewed as a reason for their relatively low effectiveness in human treatments, compared to their promising results in animals.	Some antioxidants can make it hard to cross the BBB. Herein lies the possibilities of using antioxidant therapies at an early stage of degenerative processes.
Innovative Drug Delivery Systems	Souped up drug delivery systems such as isolated nanoparticles and liposomal programs are in consideration to provide antioxidants to the brain.	These systems can improve the ability of the compound to cross the BBB better, increase the concentration of the drugs to the target site, and minimize side effects at systemic level. Further, management practices to implement genome-based therapies are under consideration in terms of combination therapies and developing the concept of personalized medicine.

Table 1.
Oxidative Stress in Neurodegenerative Disease and releated articles.

conditions. Additionally, the Nrf2 pathway, which received abundant attention from researchers recently, is shown to be a key pathway for the update of the therapeutic approach and development of the treatment. The paper discusses the main points of the literature in general, and it has updated info about how oxidative stress influences neurodegenerative diseases. Oxidative stress mechanisms partaking in neurodegeneration have been studied carefully on the molecular level with many discoveries made in understating this process. The creation of advanced therapeutic modalities according to mechanistic particular targets, such as alteration of the Nrf2 pathway, may turn out to be new strategies in the treatment of these diseases. The author stresses that the efforts should be directed at the activation of clinical research to uncover the underlying mechanism of oxidative stress all in connection with the psychological processes of neurodegenerative and give leeway to advanced treatment options.

Sajad, N; Ali, R.; Hassan, P.; and Ganai, A B. have authored an article titled "Oxidative Stress in Neurodegenerative Diseases" that thoroughly scrutinizes the part that oxidative stress plays on neurodegenerative diseases and the contribution that it accords to neurons. In this study, an in-deep overview of how the oxidative process influences the disease development in cases such as Parkinson's, Alzheimer's, and Amyotrophic lateral sclerosis is taken. The definition of oxidative stress in simple terms is a situation when oxidative damage exceeds the ability of the cells to counteract this damage. Deca is a name for the condition where it creates a balance between the synthesis of the free radicals and a cell's ability to fend off these radicals [5].

ROS are harmful molecules with the particular ability to affect the brain; why does this happen? It is simply because the brain is constantly consuming a lot of oxygen. The oxidative stress is recognized as a core event in the pathogenesis of Parkinson's disease and other severe ailments such as Alzheimer's disease and amyotrophic lateral sclerosis. These neurological disorders are distinguished by the presence of neuron (nerve cell) degeneration accompanied by cognitive impairment (memory, psychological, and sensory function). Tissue and cellular injury due to the increase of free radicals can damage lipids, proteins, and DNA, which are considered cellular molecules. This phenomenon, when left unchecked, may result in the breakdown of cells and necrosis, or cell death by apoptosis. One of the main tasks in the battle against oxidative stress can be the identification of the pathways, which are essential for the production as well as for the defense of free radicals. This understanding of the key mechanisms might lead to the development of a new and pharmacological approach to retarding or even stopping the degeneration of neurons.

This study significantly expounds an overview of the existing research and evidence that contradicts the relationship between oxidative stress and neurodegenerative diseases. It presents substantial findings on the mechanisms behind oxygen starvation and cell death, which act through neurodegenerative diseases. Furthermore, the fundamental mechanisms of free radical-based damage, overviewing how these lead to neuronal demise, are explained. The assessment of currently utilized and the prospective electron transfer-targeting treatments give us powerful tools to deal with these health problems. In this case, pharmacological intervention to increase the production of antioxidant strategies, as well as enzyme production that regulates the creation of free radicals, commonly occurs. The article reveals that more studies are essential to cutting the consequences of oxidative stress on the process that leads to neuronal deficit diseases. Also, it says that this understanding of the molecular mechanism of the process will enable the generation of effective treatment methods ultimately.

Oxidative stress (OS) as the underlying cause for the progression of neurodegenerative diseases has been extensively studied and has captured much attention over the last few decades. Particularly, the mechanism is responsible for the development of conditions such as amyotrophic lateral sclerosis (ALS), Parkinson's disease (PD), and Alzheimer's disease (AD). Niedzielska and col. have performed a detailed assessment of the contribution of oxidative stress to such conditions, with the vivid increase of stress biomarkers and the degeneration of antioxidant defense in the brain and other central and peripheral ailments tissue being one of the essential aspects. Through the collation of clinical trials and animal model findings, the research illustrates a complex image of free radical-mediated damage and its role in neurodegenerative pathologies. Moving on, Niedzielska et al. also explore drug therapy interventions targeting lower OS through oxidative balance metabolic pathways. The review is a reminder of the critical level of serious pharmacotherapy that needs further studies and the vision of the future of targeting oxidative stress to address neurodegenerative diseases. The approach described above not only helps to appreciate the intricacies as well as the limitations of the oxidative stress theory but also serves another purpose of providing a driving force for research efforts. It is believed that more effective and targeted treatments will be developed if the progression of the neurodegenerative conditions can be either reversed or slowed down [6].

The importance of oxidative stress in neurodegenerative disorders them fore the core one among the most significant fields of study, as demonstrated by Chen et al., in their comprehensive review. The authors describe free radicals such as superoxide anion (O_2^-), hydroxyl radical (OH−), and other reactive oxygen species like hydrogen peroxide (H_2O_2), which are produced as species naturally due to respiration that occurs in the body and from exposure to toxic environmental stressors such as UV light and ionizing radiation. Although ROS themselves serve as a vital intracellular communicator at lower concentrations, their ability to undergo oxidative disassembling of the DNA, lipids, and/or proteins is remarkable at higher concentrations developing necrotic as well as apoptotic cell-death mechanisms. Oxidative stress is an imbalance between the production of reactive oxygen species and the level of antioxidants to defend cells in the body; this is a defining trait of the group of neurodegenerative diseases such as Alzheimer's, Parkinson's, and amyotrophic lateral sclerosis. The development of these diseases indicates a sequence of events that includes an accumulation of protein aggregates and metal ion dysfunction. These features are usually linked to oxidative stress that has been closely studied. Simultaneously, this report not only summarizes the current knowledge about these mechanisms but also covers existing treatment strategies that function through diminishing oxidative stress *via* Nrf2-mediated transcription factor, which in turn will serve an important role in developing future treatment of such life-threatening diseases [4].

The review paper written by Li et al. [7] extensively highlights oxidative stress's role in neurodegenerative diseases and shows a double-edged role played by reactive oxygen species (ROS) in the body's processes. Concentrated steadily through aerobic respiratory phosphorylation, they are often needed at these physiologically low to medium concentrations. Nevertheless, the ROS' overproduction leads to oxidative stress, which is the biggest enemy of the CNS system, especially the CNS being the part of the body with the strongest oxygen demand but without enough antioxidative defenses and the non-replicative neurons. According to Li et al. the authors describe how oxidative stress is tightly related to several neurodegenerative disorders, suggesting the exceptional sensitivity of the central nervous system to it.

Along the lines of this theme, the evaluation also goes beyond the established neurodegenerative disease models, with which the toxic effects of chemotherapy are considered in terms of not only the brain and the nervous system but also the peripheral nervous system. ROS play a central role in chemotherapy-related neurotoxicities, and there is evidence provided that they are the ones that bring about the oxidative stress changes posttreatment besides neurodegeneration and therapeutic neuropathies. In addition, as Li et al. comment, the antioxidants can be prepared in such a way that they will be termed neuroprotective agents. An overview of the current insights on the intracellular redox signaling mechanism and its association with disease is provided by their review. Moreover, an assessment of the possible and limited effectiveness of antioxidants used in medicinal therapy is made in their review. This assessment is undoubtedly a functional precursor to further studies on the use of antioxidants that dodge oxidative stress, which leads to neurodegeneration in such diseases as Alzheimer's disease, as well as in neurological damage caused by chemotherapeutic agents [7].

In their systematic review, Kim et al. detailed the connection of oxidative stress to the beginning and the progression of neurodegenerative disorders, which was primarily among the topics widely studied due to the rapidly aging population. Oxidant stress is the term given to the conformation that is created in the environment by excessive production of ROS or a failed antioxidant system. The brain, which consumes a lot of oxygen and also contains many lipid cells (these are more susceptible to peroxidation), becomes more vulnerable to the effects of the free radicals. As presented by Kim et al.'s analysis, oxidative stress is a major facet of mixed pathophysiology of the most common neurodegenerative diseases like Alzheimer's disease (AD) and Parkinson's disease (PD). Oxidation or free radicals are therefore considered to be the causes of these diseases, which function by damaging cellular structures and disruptions of normal functions. Even though the use of Antioxidants in the hopes of neutralizing this oxidant damage, in theory, is promising, the findings of a complex mechanism between antioxidants and the deteriorating pathology of the brain in the review called into question the results of clinical studies. To move on, Kim et al. publication, which is aimed at the *in vivo* measurement of oxidative stress markers' different methods, is discussed; as a result, we get acquainted with the technologies' modern possibilities and limitations. In addition to this, the authors also mention a few research directions of antioxidant therapies that include effective and more directed strategies for applying antioxidants in the prevention and cure of neurodegenerative diseases [8].

3. New diseases linked to oxidative stress and redox imbalance

Oxidative stress or the overload of ROS in tissue that the body can handle or neutralize or make up for the damage it causes has been attributed to the etiology of many diseases. As described earlier, oxidative stress has been observed to cause serious neurodegeneration, and neurodegenerative disorders like Alzheimer's disease (AD) and Parkinson's disease (PD) are extremely prevalent. Nevertheless, subsequent studies have described the role of ROS, free radicals, and antioxidant defenses as the central mechanistic axis playing a significant, even causative, role in a growing list of more modern diseases and pathological states that include premature brain aging, tauopathies, and other new forms of neurodegenerative disorders. In this essay, new diseases related to oxidative stress will be described concerning their causes, pathophysiology, as well as possible treatments. Aging of the brain is a normal physiological process through which there is a progressive loss of functionality and disorganization of parts of the brain which is responsible for our thinking capabilities [9]. But certain conditions can cause this process to be faster than the normal rate and consequently cause early aging of our brains. Oxidative stress has been well recognized as a major factor contributing to the rapid aging of the brain based on the high oxygen requirement of the brain tissues and the presence of high lipid content.

These characteristics make neurons one of the more vulnerable cell types concerning oxidative stress. Among all structures, it is the mitochondria that are the main producers of ROS. It has also been demonstrated that there is a gradual decline in the functional ability of mitochondria over time, which results in increased generation of ROS and damage to both the MT DNA and proteins. This damage depletes energy in the cell and affects the mitochondria, thus forming a cycle that is very destructive to the cell and shortening its lifespan. The stressor always enhances the formation of injured and misfolded proteins to augment the overall oxidative stress. This in the brain can bring about the formation of aggregates through interference with cellular functions and bring about the loss of neurons. ROS generated continuously in the brain cause microglial activation and chronic neuroinflammation due to the relentless changes in their phenotypes [10].

This inflammation not only amplifies neuronal damage and enhances the aging process, but it has also altered the innate immune system and the individual's susceptibility to infections. Hormones including resveratrol and coenzyme Q10 have been found to possess the ability to prevent oxidation and promote mitochondrial biosynthesis. These tauopathies are neurodegenerative diseases that may lead to a progressive cognitive impairment of the affected patients due to the presence of tau protein in the cells. AD and FTD are familiar examples of neurodegenerative diseases most of whose early family members present as dementia. Oxidative stress is another crucial variable that affects phosphorylation and redistribution of tau protein in the cell. Reports also show that oxidative stress upregulates kinases involved in signal transduction pathways and downregulates phosphatases involved in the dephosphorylation of tau. The deposition of excessive phosphate will cause the aggregation of hyperphosphorylated tau proteins and form neurofibrillary tangles, a key pathological feature of AD and other tauopathies.

Usually, tau protein helps accumulate and stabilize microtubules in neurons, which are the nerve cells in the brain. In its hyperphosphorylated state, tau comes off the microtubules, and the structure becomes compromised; the axonal transport that is necessary for proper neuronal health is disrupted as well. The activated microglia secrete pro-inflammatory cytokines, as well as release ROS, therefore enhancing tau pathology and neuronal injury. New research indicates the possibility of controlling specific kinases attributable to tau phosphorylation to halt the formation of the neurofibrillary tangle [11]. This may suggest that these inhibitors can help to treat tauopathies at some stage because they will have slowed the rate of the diseases. Curcumin and vitamin E are among the several compounds that have been identified to have the potential to minimize oxidative stress or tau pathology, making tauopathies a probable therapeutic target.

They identified that PD is the degeneration neurons of the dopaminergic substantia nigra, which cause motor disorders. Dopaminergic neuronal toxicity and loss is one of the most important features of PD and oxidative stress that contributes to this process in several ways. For example, PINK1 and PARKIN genes; their mutated versions are known to damage mitochondrial function and, consequently, raise ROS production levels. The toxic gain of function is due to oxidative stress, which leads to misfolding of α-synuclein into insoluble aggregates known as Lewy bodies. Dopamine metabolism also produces ROS within cells, and these free radicals can also damage the dopaminergic neurons [9]. This could be due to the capability of some compounds, such as coenzyme Q10 and creatine, to elevate the mitochondrial form of the antioxidants. The toxicity, accumulation, and formation of insoluble Lewy bodies through the aggregation of α-synuclein protein are some of the major factors observed in the pathogenesis of PD, and targeting α-synuclein with small molecules or immunotherapy may represent a potential therapeutic avenue. ALS leads to the death of the motor neurons of the body. Oxidative stress resulting from compromised mitochondrial activity and excessive release of excitatory amino acids is a key feature of the disease in ALS. Promoting antioxidant protection and conserving mitochondrial membrane potential could be useful. In HD, people develop the mutation in the huntingtin gene, which results in the accumulation of the mutated huntingtin protein. Because oxidative stress decompensates and aggravates both protein aggregation and neuronal cell death in HD. Low-energy therapy extends lifespan in fly models of HD, and therapies that address oxidative stress and protein aggregation may be effective for HD.

MS is an inflammatory autoimmune disease of the central nervous system that results in the destruction of the protective sheath (myelin) around neurons. The researchers showed that many disease processes occurring in multiple sclerosis may be explained thanks to oxidative stress. Corticosteroids, another immunomodulatory therapy under investigation for the treatment of MS, is IL-2. Reactive oxygen species and oxidation damage of cellular components are key mechanisms that drive the development and progression of many neurodegenerative diseases and premature aging of the brain [10]. This is because understanding exactly how oxidative stress and antioxidants affect the brain and hence neuronal well-being is a necessary step towards providing the right kind of treatment. Although all the antioxidant therapies have demonstrated benefits the drawbacks are related to bioavailability, crossing the BBB, and finding the right dosage. Consequently, the issues have negatively impacted health care as it needs to contain the above challenges for the following reasons: Advanced delivery systems, personalized medicine approaches, and combinational therapies.

4. Discussion

Investigation of oxidative stress as a central feature of neurodegeneration paves the way for therapeutic trials that attempt to rebuild oxidative stress balance, resulting in a diminished accompaniment effect of neurodegenerative diseases. The aging process may include several harmful biochemical processes; the understanding of these biological events, as well as the control over them, has been the key point for the development of several successfully treated diseases. Meanwhile, these two findings highlight the need to investigate further into the matter, unraveling the factors that may have led to such a result. At the same time, the ongoing recovery strategies might also require revamping. Owing to antioxidants, various experiments in test tubes and animals found oxidative damage decreasing and the progression of neurodegenerative disorders slowing down. Nevertheless, the reseeds of such findings in the clinical trial space are just inconsistent and do not necessarily reproduce similar positive outcomes.

This may be based on certain factors like pharmacokinetics of antioxidant compounds, which could influence the bioavailability of their crossing of BBB. One of the major hindrances for many antioxidants is the BBB, which acts as a signatory barrier, leading to a lower concentration of these agents reaching the action site. However, the nutritional intervention timing when providing nutrition as intervention complements the disease progression is equally important. The effectiveness of antioxidant therapy could be higher if it was somehow utilized during the early onset of the disease before the damage is irreversible. In doing so, the possible moment at which antioxidant therapy can be most effective may just pass by. To surmount the restraints obstructing the delivery of antioxidants to the brain, novel drug delivery systems including, nanosystems, liposomal encapsulation, and molecular transporters have been researched. They can improve the crossing of the BBB, pass the brains on the specific area, and, with that the concentration in the site of interest, reduce systemic side effects.

5. Conclusion

The path that starts from grasping the fundamental oxidative stress mechanism in neurodegenerative diseases and continues till the development of successful antioxidant tablets has seen many obstacles along the way. Nevertheless, the future throws some challenges on the way to the precise regulation of the delivery of antioxidants; the deeper monitoring of the molecular mechanisms of the disease accompanied by the customization of treatment creates the guarantees that the future will not just contain the treatment of the neurodegenerative diseases, but will rather allow defeating them. With the advancement in research, it is expected that the coming discoveries will be turned into trajectory therapies that are more efficient and targeted, and this will relieve the problem of these devastating diseases. Investigations into oxidative stress and its implications for neurodegeneration have haunted biomedical research with the problems and opportunities that they entail to strive towards effective therapies. This review has evaluated in a critical way the function of oxidative stress in the pathophysiology of illnesses including Parkinson's infection, Alzheimer's disease, Huntington's disease, and multiple sclerosis, bringing to light the dual function of ROS as both detrimental agents and molecular signaling molecules.

Although antioxidant therapies worked well during the initial animal studies, they still fail to provide a positive outcome in human clinical conditions, no matter how promising the results seem, which still raises the question of the translatability of laboratory findings into the proposed therapeutic successes. One core challenge that prevents antioxidant application in the therapy of neurodegenerative disorders is the lack of capability of these chemicals to cross the blood–brain barrier. The advent of novel drug administration strategies, nanoparticles, and liposomal carriers, among others, is one promising solution to difficulties in the central nervous system delivery, as well as to achieve therapeutic efficacy. The important element of antioxidant administration is the time of action and timing matters. Actually, in the initial stage, the damage is at its peak and the hope for faster repair is more likely. This is accordingly contributing to early treatment whereby it prevents or minimizes the effects of degeneration of the brain, which gets worse with time. The activation of the body's antioxidant defenses, especially by cueing the Nrf2 pathway, over compounds antioxidant supplements is far more economical. The complexity of neurodegenerative diseases including several dysfunctioning pathways means that treatments are optimally multidirectional. Looking forward to future research, the goal should be implementing multifeatured methods that unite the antioxidant, anti-inflammatory, and neuroprotective activities. An advanced, more broad-minded solution should be applied to aid the process of antioxidant therapy creation and its implementation. It contains two parts: one of them involves giving more support to customized medicines based on people's particular genetic backgrounds, environmental exposures, and disease stages. Furthermore, it usually requires larger and better trials set out to give guidance about what doses and conditions fit for antioxidants to work better. In general, with the road to developing effective antioxidant therapies for neurodegenerative diseases marred with more failures than gains, the prospect of their role in treating such diseases still holds sizable value. Improved drug delivery technologies, which continuous advancements bring, better disease mechanisms knowledge, and individualized treatment, will help to overcome oxidative stress obstacles in the neurodegeneration context. The devotion to developing these treatment options is no doubt not a simple path but a crucial one as it gives hope for patients who are struggling with these disabling conditions.

Author details

Yasemin Gündüztepe
Faculty of Medicine, Department of Biochemistry, Gazi University, Ankara, Turkey

*Address all correspondence to: yasemin.gunduztepe@gazi.edu.tr

References

[1] Carvalho AN, Firuzi O, Gama MJ, Horssen JV, Saso L. Oxidative stress and antioxidants in neurological diseases: Is there still hope? Current Drug Targets. 2017;**18**(6):705-718

[2] Domanskyi A, Parlato R. Oxidative stress in neurodegenerative diseases. Antioxidants (BASEL). 2022;**11**(3):1-504

[3] Olufunmilayo EO, Gerke-Duncan MB, Damian Holsinger RM. Oxidative stress and antioxidants in neurodegenerative disorders. Antioxidants. 2023;**12**(2):517

[4] Chen X, Guo C, Kong J. Oxidative stress in neurodegenerative diseases. Neural Regeneration Research. 2012;**7**(5):376-385

[5] Sajaad N, Ali R, Hassan S, Ganai AB. Oxidative stress in neurodegenerative diseases. International Conference of Management, Technology And Engineering. 2018;**8**(7):508

[6] Niedzielska E, Smaga I, Gawlik M, et al. Oxidative stress in neurodegenerative diseases. Molecular Neurobiology. 2016;**53**:4094-4125

[7] Li J, Wuliji O, Li W, Jiang Z-G, Ghanbari HA. Oxidative stress and neurodegenerative disorders. International Journal of Molecular Sciences. 2013;**14**(12):24438-24475

[8] Kim GH, Kim JE, Rhie SJ, Yoon S. The role of oxidative stress in neurodegenerative diseases. Experimental Neurobiology. 2015;**24**(4):325-340

[9] Kıran T, Otlu O, Karabulut BA. Oxidative stress and antioxidants in health and disease. Journal of Laboratory Medicine. 2023;**47**(1):1-11

[10] Rey F, Berardo C, Maghraby E, Mauri A, Messa L, Esposito L, et al. Redox imbalance in neurological disorders in adults and children. Antioxidants. 2023;**12**:965

[11] Li Z, Xu D, Li X, Deng Y, Li C. Redox imbalance in chronic inflammatory diseases. BioMed Research International. 2022;**202**:9813486

Chapter 8

Oxidative Stress and Antioxidant Interventions in Type 2 Diabetes

Aamir Sohail, Muhammad Murtaza Hasnain, Muhammad Ehsan Ul Haq, Irum Nasir, Raza Sufyan, Mehroz Khan and Imran Ullah

Abstract

The pathophysiological basis of the onset of type 2 diabetes mellitus (T2DM) focuses on oxidative stress that is precipitated due to an imbalance between reactive oxygen species and antioxidants. Gradual and chronic hyperglycemia, as seen in T2DM, produces ROS, which stimulates cellular damage, inflammation, and insulin resistance. The current chapter will focus on endogenous and exogenous antioxidant strategies to ameliorate oxidative stress and its complications in T2DM. Of course, naturally occurring oxidatively damaging effects due to endogenous antioxidants like non-enzymes, glutathione, alpha-lipoic acid, and enzymes-for instance, superoxide dismutase and catalase-. Exogenous antioxidants in the diet can provide further protection or as supplements since products like polyphenols, flavonoids, and vitamins, among other products, have effectively reduced oxidative stress and improved insulin sensitivity. Additionally, PTP1B is a prime drug target since the protein dephosphorylates insulin receptors and supports insulin resistance. Research has shown that blockage of PTP1B increases glucose uptake and stimulates insulin signaling. Therefore, research on such interventions should be continued for T2DM treatment, especially concerning antioxidant therapy.

Keywords: oxidative stress, type 2 diabetes mellitus, antioxidants, protein tyrosine phosphatase 1B (PTP1B), reactive oxygen species (ROS)

1. Introduction

Diabetes is a long-term metabolic disorder that is characterized by high blood sugar levels, which can cause serious complications if they are not treated or not properly managed. The main causes of diabetes include insulin resistance, reduced insulin secretion, and genetic predisposition [1]. There are three main types of diabetes: type 1, type 2 (T2DM), and gestational diabetes. An autoimmune disease, type 1 diabetes usually develops in youth or young puberty, with the pancreas incapable of synthesizing insulin. T2DM, accounting for most cases and normally associated with old age, is insulin resistance combined with reduced insulin secretion, the other vocal about its causes being numerous lifestyle factors, mainly obesity [1]. Recognizing several types of diabetes is fundamental to diagnosing what is going on and applying appropriate treatment and management strategies to minimize complications.

1.1 Background on type 2 diabetes mellitus

The common symptoms of the condition include excess glucose and lack of insulin in the blood, which is also known as hyperglycemia and insulinopenia [2]. Insufficient insulin production or its inadequate function in the body can cause imbalances in the metabolism of fats, carbohydrates, and proteins [2]. This causes a subsequent increase in insulin resistance in the liver, adipose, and skeletal tissues and eventually leads to T2DM, the hallmark of which is hyperglycemia because of impaired insulin secretion and pancreatic β cell dysfunction [3, 4].

1.2 Global burden of T2DM

The global burden of T2DM has gained significant concern over the past decades and is anticipated to affect over 640 million individuals by 2040. Over 21.6 million new cases were reported in 2019 worldwide, an incidence increase of special concern in high-Social Development Index (SDI) regions, such as Central Asia and Southern Sub-Saharan Africa [5, 6]. Estimates expect that global diabetes prevalence will rise to 644 million by 2030 and 783 million by 2045 [7]. The age-standardized incidence rate increased from 2022 to 2024, especially in high-income countries. Similarly, T2DM-related deaths and disability-adjusted life years (DALYs) also increased dramatically over this period. However, the age-standardized death rate rose from 16.69 deaths per 100,000 in 1990 to 18.49 deaths per 100,000 in 2019, with marked increases in low and middle SDI regions [8]. Thus, the immediate introduction of holistic public health strategies targeting the prevention, early detection, and treatment of T2DM is required to diminish its worldwide burden [9, 10].

1.3 Role of oxidative stress in diabetes pathogenesis

Free radicals that are highly reactive species can be generated in normal cellular metabolism or can be entered exogenously. Cell organelles (mitochondria, peroxisomes, and endoplasmic reticulum) and external sources (high temperatures, environmental pollutants, cooking oil, and radiation) produce reactive molecules [11]. These include oxidative radicals, reactive oxygen species, and other biomolecule-damaging free radicals, which can react with lipids, proteins, and DNA through diverse reactions [11, 12]. Antioxidants counteract the effects of free radicals produced by metabolic processes. The cell is said to be under oxidative stress when the production of reactive molecules like ROS, RNS, or RSS is high. These molecules accumulate in the body due to the failure of the antioxidant defense system. In several chronic and degenerative disorders, oxidative stress participates in disease progression [13].

Oxidative stress also has a crucial role in diabetes pathogenesis. Chronic hyperglycemia produces ROS, which results in DM. Excessive ROS generation leads to excessive ROS accumulation, which is linked with mitochondrial dysfunction and can lead to IR and T2DM (**Figure 1**) [14, 15]. ROS species can be responsible for developing mitochondrial impairment by distorting the protein function by oxidizing on Methionine and Cysteine residues, so the protein will lose proper functioning

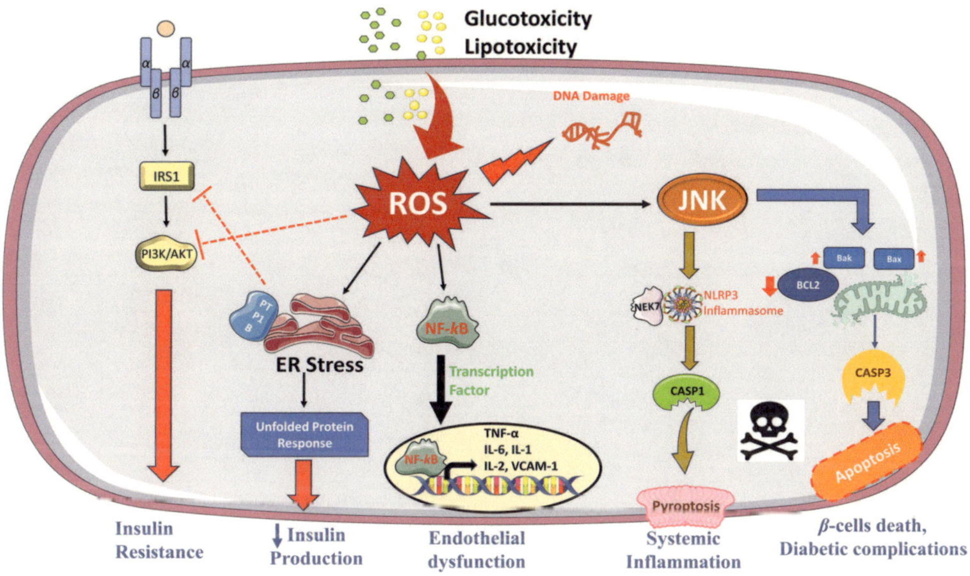

Figure 1.
Molecular mechanisms of oxidative stress in T2DM.

and cause cell death [16]. Aggregation of ROS in cells leads to activation of PKC, polyol pathway activation, modification of Growth factors expression, initiation of hexosamine pathway, and enhanced production of glycation product [17]. Defective mitochondria with elevated oxidative stress led to the activation of the mitophagy process [18].

1.4 Overview of protein tyrosine phosphatase 1B (PTP1B) and insulin signaling

The insulin signaling starts when insulin attaches to the insulin receptors (INSR), which activates a series of pathways in succession. Insulin attaches to the INSR and activates it. Activated INSR phosphorylates downstream scaffold proteins of the insulin signaling cascade known as insulin receptor substrates (IRS1 and IRS2) (**Figure 2**) [19]. Phosphatidylinositol 3-kinase (PI3K) p85 subunit activation is associated with the phosphorylation of substrates IRS1 and IRS2. This leads to AKT (PKB) activation, further phosphorylating the downstream proteins. These activated proteins work with the small GTPase RAB10 to speed up the glucose transporter GLUT4 vesicle localization to the cell surface and encourage insulin action and glucose storage [20].

Several mediators regulate the insulin-signaling pathway, and tyrosine phosphorylation has long been known to control several cellular processes. One of the best-studied tyrosine phosphatases, PTP1B, plays a crucial regulatory role in various signaling cascades. PTP1B, with a molecular weight of 49,967 Da, is encoded by the PTPN1 gene. It modulates insulin and leptin receptors directly, making it an ideal therapeutic target for T2DM and related diseases [21]. It is ubiquitously present in the liver, muscles, adipose tissue, and brain, particularly on the cytoplasmic surface of the endoplasmic reticulum, binding through a hydrophobic sequence of 35 amino acids at the C-terminal side [22]. PTP1B expression is regulated by elevated levels of free fatty acids (FFAs), incredibly saturated FFAs, which activate the JNK–NF-κB pathway, leading to elevated PTP1B expression [23].

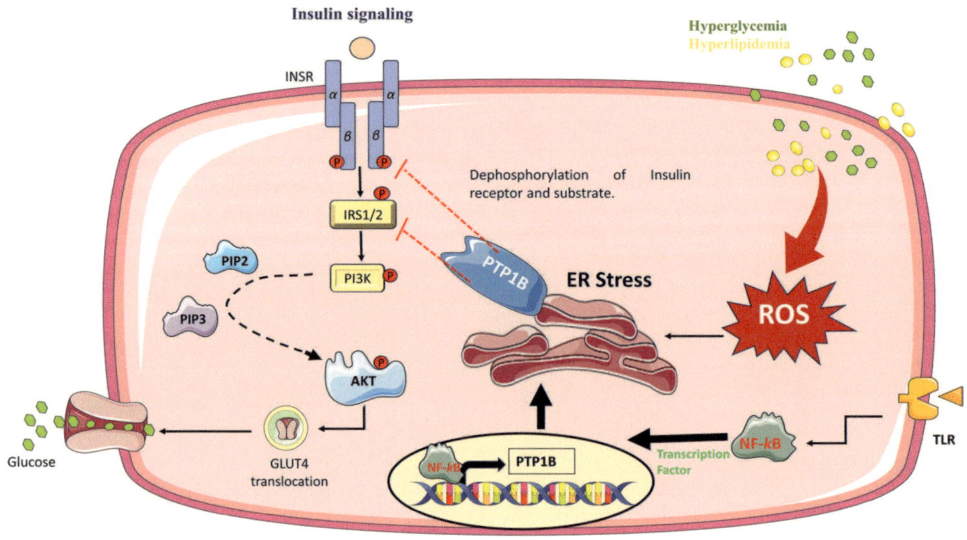

Figure 2.
ROS impact on insulin signaling cascade.

Additionally, inflammatory markers such as TNFα, IL-6, and IL-1β augment PTP1B expression by activating the NF-κB pathway [24–27]. ER stress also enhances PTP1B expression by stimulating the ROS-NF-κB pathway [28]. PTP1B, a class of hydrolases, inhibits insulin signaling through the tyrosine dephosphorylation of IRS-1 and INSR-β, thereby inhibiting the activation of downstream mediators of the PI3K-AKT pathway and contributing to cytokine production that controls inflammation [29–31]. It also negatively regulates the leptin signaling pathway by dephosphorylating STAT3 and JAK2 and is overexpressed in various tissues, including the liver, pancreas, adipose tissues, and muscles, in conditions of obesity and T2DM [32]. PTP1B regulates the inflammatory response in the brain, macrophages, and other organs, affecting hepatocyte apoptosis and regeneration abilities [33]. It contributes to the negative regulation of the leptin pathway by dephosphorylating JAK2 downstream of the LR [34]. Overexpression of PTP1B is linked to dyslipidemia in T2DM, downregulating adipogenic markers such as PPARγ, FAS, SREBP1c, and lipoprotein lipase (LPL), while its knockdown upregulates these markers [35]. In the liver, PTP1B expression is associated with elevated gluconeogenesis, hepatic lipogenesis, aberrant insulin signaling, and hepatic inflammation, indicating that PTP1B negatively regulates insulin signaling and is linked to poor prognosis in T2DM [36].

1.5 Importance of antioxidant interventions

Antioxidants are compounds that inhibit the production of free radicals. These compounds play a significant role in preventing damage from oxidative stress. They create a barrier between oxidative stress and cellular damage. Antioxidants are two major types: endogenous and exogenous antioxidants. Endogenous antioxidants include enzymatic (superoxide dismutase, catalase, and glutathione systems) and non-enzymatic (glutathione, ferritins, melatonin, bilirubin, thiols, uric acid, ubiquinone, and transferrin) antioxidants [37]. Exogenous antioxidants include flavonoids, genistein, carotenoids, phenolic acids, and vitamins [38]. All antioxidants work through unique mechanisms and scavenge ROS/RNS and chelate metals that can create free radicals. Antioxidants possess ROS/RNS scavenging and free radical neutralizing properties, making them a potential treatment option for various diseases.

Antioxidants prevent free radical-induced tissue damage by inhibiting the formation of radicals, scavenging them, or promoting their decomposition [39]. This is an emerging research area with various studies identifying the role of antioxidants in disease prevention and therapeutics. For instance, oxidative/nitrosative stress-based chronic inflammation is the leading cause of inflammatory bowel disease (IBD), and different antioxidants, including medications, hormones, and enzymes, are used in its therapy [40]. Statins, which have antioxidant effects, are used not only in cardiovascular diseases and atherosclerotic therapies but also to reduce the risk of colorectal cancer in IBD patients and the need for biological or immunosuppressive treatments [41, 42]. Additionally, angiotensin-converting enzyme inhibitors (ACEIs) like telmisartan, captopril, and valsartan exhibit antioxidant and anti-inflammatory properties by lowering inflammatory proteins such as TNF-α, MPO, MDA, and iNOS [42, 43]. Telmisartan also enhances intestinal defense mechanisms against free radicals by increasing GSH and TAC levels and the activity of antioxidant enzymes such as SOD and GPx [44]. Other antioxidants like melatonin, N-acetylcysteine, modified superoxide dismutase, and propionyl-l-carnitine have been shown to reduce ROS levels in ulcerative colitis and Crohn's disease [45–48]. Plant-based antioxidants, such as polyphenols and resveratrol, are also used as complementary medicine in IBD to reduce lipid peroxidation and enhance antioxidant protection by increasing GPx enzyme activity [49, 50].

Oxidative stress significantly contributes to the development of cardiovascular diseases, including atherosclerosis, heart failure, cardiac arrhythmia, and ischemia-reperfusion injury. Several therapeutic options target these conditions, including preclinical studies showing that both astaxanthin and vitamin A protect against ischemia/reperfusion injury and thrombotic diseases [51, 52]. Astaxanthin reduces blood pressure in hypertensive rats and decreases serum lipid peroxidation while increasing SOD levels [46]. Vitamin C detoxifies ROS, controls endothelial cell proliferation and apoptosis, and mediates vasodilation, which is crucial in cardiovascular diseases [53]. Vitamin E inhibits superoxide production by impairing NOX enzyme assembly, reducing the risk of coronary heart disease and cardiovascular complications [54]. Experimental antioxidants, such as allopurinol, an XO inhibitor, and GKT137831, a NOX inhibitor, have shown promising results against oxidative stress-related cardiovascular diseases [55, 56]. miRNAs are also being explored as therapeutic options; for example, inhibition of miRNA-133 can prevent endothelial dysfunction by targeting NOS, while miRNA-92a promotes endothelial cell proliferation and angiogenesis, reducing plaque inflammation and increasing plaque stability [57]. In sepsis, early antioxidant intervention to remove excess ROS/RNS has been beneficial, with the de novo-designed Cu-SAzyme showing exceptional SOD-like activity to neutralize superoxide and block the inflammatory response during sepsis efficiently [58]. Additionally, antioxidants play a significant role in metabolic disorders like T2DM, where supplementation of compounds like melatonin, Viscosol, and vitamins E and C improves oxidative stress and insulin signaling, reducing complications associated with the disease [59, 60].

1.6 Objective of the chapter

This study aims to investigate how free radicals and oxidative stress contribute to the development of T2DM. Additionally, we will analyze the effects of oxidative stress on PTP1B and insulin resistance. Finally, we will discuss new antioxidant approaches and potential targets for treating T2DM.

2. Oxidative stress and T2DM

2.1 Molecular mechanisms of oxidative stress

Oxidative stress is the imbalance between reactive oxygen species and antioxidants. It plays a vital role in the pathogenesis of type 1 and 2 diabetes mellitus. The disruption of redox equilibrium, resulting in cellular malfunction, fosters a pro-oxidative milieu where the equilibrium between the generation of ROS or RNS and their removal in the organism is compromised. This state, known as oxidative stress, culminates in cellular and tissue impairment, precipitating the onset of metabolic disorders [61]. Consequently, numerous investigations in preventing and managing metabolic ailments have explored the interplay between oxidative stress and the advancement of such conditions. The defensive system includes different enzymes such as superoxide dismutase (SOD), Catalase (CAT).

ROS, hydroperoxyl, superoxide, hydrogen peroxide, hydroxyl radicals, and reactive nitrogen species (RNS: peroxynitrite) are examples of active derivatives of oxygen and nitrogen molecules that are free radicals [61]. Copper and iron (ferric) are heavy metal derivatives with free radical characteristics. Due to unpaired electrons in their outermost molecule, these hyperactive components can attach and alter other biomolecules [61–63]. They can oxidize nucleic acids, proteins, and lipids, producing harmful byproducts that impair tissue function [64]. They alter the structures of biological molecules and even break them. DNA breakage is one well-known consequence of oxidative stress, which impacts gene expression and cell viability [61]. In addition to their direct harmful effects, free radicals can also cause indirect cell damage through the activation of various stress-sensitive intracellular signaling pathways, including sorbitol synthesis, JNK/SAPK (stress-activated protein kinase/c-Jun NH (2)-terminal kinase), AGE/RAGE (advanced glycation end product/receptor for AGE) interactions, and Nf-κb (nuclear factor kappa b) [65]. Oxidative stress is critical in the pathogenesis of diabetic problems, including lipid peroxidation, DNA damage, and mitochondrial dysfunction [65].

2.2 ROS and cellular damage

ROS damage the cells by direct DNA damage or by activation of signaling pathways. It significantly disrupts normal cellular function and can lead to inflammation, tissue injury, and organ dysfunction. Free radicals damage the cellular protein through various mechanisms, including amino acid modification, cross-linkage formation due to lipid peroxidation, and protein fragmentation. In proteins, methionine, cysteine, arginine, and histidine are the most vulnerable to oxidation [66]. Enzyme proteolysis is more likely in proteins already damaged by free radicals. Oxidation of protein products affects the activity of enzymes, receptors, and the passage of molecules across membranes. Highly reactive groups in oxidatively damaged protein products can induce membrane damage and hinder cellular functions [67]. Peroxyl radicals, a form of free radicals, are thought to be responsible for protein oxidization.

ROS can generate carbonyls and other amino acid changes by destroying proteins and producing methionine sulfoxide and protein peroxide. Protein oxidation impacts several aspects, including signaling pathways, enzyme function, heat stability, and proteolysis susceptibility [68].

Cell membrane-bound polyunsaturated fatty acids undergo lipid peroxidation through a radical chain reaction. This process is believed to be initiated by hydroxyl radicals, which abstract hydrogen atoms, creating lipid radicals and diene conjugates [69]. When oxygen is incorporated, a highly reactive peroxyl radical forms and attacks another fatty acid, producing lipid hydroperoxide (LOOH) and a new radical. This sequence amplifies lipid peroxidation [70]. Several byproducts, such as alkanes, malonaldehyde, and isoprostanes, are generated during this process. These compounds have been identified as biomarkers of lipid peroxidation in various conditions, including diabetes, ischemia-reperfusion injury, and neurodegenerative disorders [71].

Oxidative DNA damage is an unavoidable result of cellular metabolism. Usually, guanine pairs with cytosine, but the most prevalent form of oxidative base damage, 8-oxo-7,8-dihydro guanine (8-oxoG), can cause it to mispair with adenine due to a conformational shift [72]. Numerous studies have conclusively shown that DNA and RNA are susceptible to oxidative stress damage. Various diseases, including aging and cancer, have been associated with DNA mutations. Oxidative DNA damage, caused by free radicals or ultraviolet radiation, results in elevated levels of oxidative nucleotides such as glycol, DTG, and 8-hydroxy-2-deoxyguanosine [73].

2.3 ROS and insulin signaling

ROS generation is a critical element of insulin signaling, yet oxidative stress can trigger various abnormal pathways that disrupt this process. In diabetes and obesity, excessive ROS production induces intracellular stress, contributing to the endoplasmic reticulum (ER) and mitochondrial stress, leading to abnormal insulin signaling and insulin resistance [74]. ROS can provoke ER stress, activating the unfolded protein response (UPR) and adversely affecting insulin production. A significant consequence of ER stress is the activation of PTP1B, which dephosphorylates insulin receptors and their substrates (INSR and IRS1/2), causing insulin resistance [75].

Additionally, abnormal ROS levels activate other kinases and phosphatases, including PTEN, PHLPP2, and MST1. PTEN plays a crucial role in negatively regulating insulin signaling by converting PIP3 back to PIP2 within the PI3K pathway [76]. PHLPP2, an essential phosphatase involved in insulin signaling and adipocyte regulation in diabetes, is activated by ROS and functions as an AKT Ser473 phosphatase. The dephosphorylation of AKT at Ser473 by PHLPP2 inhibits insulin signaling [77]. Moreover, ROS in tissue activates MST1, dephosphorylating AKT, impeding insulin signaling. MST1 is further implicated in beta cell death by activating JNK-mediated inflammatory and apoptotic pathways [78]. ROS-induced mitochondrial stress activates various stress-sensitive kinases, such as JNK1 and ERK1/MAPK. JNK1 activation is associated with obesity-induced insulin resistance and reduced compensatory insulin secretion. Additionally, JNK1 activates the NLRP3 inflammasome and the Cytoc/CASP9 apoptosome, leading to pyroptosis and apoptosis in the liver, pancreas, adipose tissue, and skeletal muscle [79]. JNK1 can also directly inhibit IRS1, contributing to the disruption of insulin signaling [80].

2.4 Oxidative stress-mediated insulin resistance

Oxidative stress plays a vital role in diabetes [81]. In T2DM, insulin resistance is caused by free circulating fatty acids, and persistent hyperglycemia causes elevated ROS production, which activates the mitogen-activated kinase inflammation pathway and NF-kB [82]. Certain pathological conditions like diabetes and obesity disturb the ER function, resulting in the misfolding of proteins [83], and ER cannot transfer unfolded protein toward the Golgi apparatus [84]. Many proteins present in ER require calcium, while depletion of calcium source due to ER disruption results in activation of many resident proteins like PTP1B, which also cause ER stress [85]. It is reported that overexpression of PTP1B can cause IR [86].

2.5 Protein tyrosine phosphatase 1B (PTP1B) and insulin resistance

In 1992, the Nobel Prize was awarded to Edmond H. Fischer and Edwin G. Krebs for their discovery that reversible covalent modification of a protein by phosphate regulates its function. Enzymes that add phosphate are phosphorylase kinases, and those that remove phosphate are phosphorylase phosphatases [22]. PTP1B, encoded by the PTPN1 gene located on chromosome 20q13.13, is a classical protein tyrosine phosphatase that regulates insulin signaling. PTP1B terminates insulin signaling by dephosphorylating the insulin receptor (IR) and insulin receptor substrate-1 (IRS-1), thus increasing insulin insensitivity and decreasing glucose uptake [87].

The PTP1B enzyme is ubiquitously present on the cytoplasmic surface of the ER, binding through a hydrophobic sequence of 35 amino acids at the C-terminal side [88]. PTP1B negatively regulates the leptin pathway by dephosphorylating Jak2 downstream of the leptin receptor (LR) [89]. Activation of PTP1B expression is regulated by elevated levels of free fatty acids (FFAs), particularly saturated FFAs. Studies have shown that inflammatory markers such as TNFα in various cell lines lead to the activation of the NF-κB pathway, resulting in enhanced expression of PTP1B [90]. Additionally, ER stress can stimulate the ROS-NF-κB pathway, leading to increased PTP1B expression, which is associated with elevated leptin and insulin resistance in muscle, liver, and adipose tissues [91, 92]. Thus, PTP1B is a primary pharmacological target for treating T2DM by downregulating insulin signaling through dephosphorylating critical insulin signaling molecules [60].

3. Antioxidant interventions in T2DM

Antioxidants are essential chemical compounds recognized for their numerous functions, such as scavenging free radicals, lowering oxidative stress, protecting the body against oxidative degeneration, and preventing the production of toxic oxidative products [93].

3.1 Endogenous antioxidants

Endogenous antioxidants help maintain the body's oxidative balance by preventing oxidation through free radicals scavenging. Endogenous antioxidants can be classified into two types based on their nature and chemical properties: enzymatic and non-enzymatic antioxidants. Enzymatic antioxidants are proteins and act as the first line of defense against oxidative stress in the body. They convert oxidative products into hydrogen peroxide (H_2O_2) and water in the availability of co-factors such as Cu, Zn, or selenium [94]. Catalase (CAT), superoxide dismutase (SOD), and glutathione peroxidase are examples of enzymatic antioxidants. Non-enzymatic antioxidants are formed within the body and are non-protein in nature. Non-enzymatic antioxidants, such as glutathione, alpha lipoic acid (LA), coenzyme Q, and ferritin, inhibit the free radical chain reaction [59]. A list of some endogenous antioxidants is given in **Table 1**.

3.2 Exogenous antioxidants

Exogenous antioxidants are given to the body through diet or medicines. Antioxidants in natural food, such as polyamines, carotenoids, flavonoids, glutathione, alpha-lipoic acid, and vitamins E, A, and C, provide substantial protection against diabetes. Research indicates that antioxidant therapy inhibits oxidative stress-induced beta-cell death, maintains beta-cell function, and decreases diabetes-associated complications. So, naturally occurring antioxidants have increased dramatically because of their natural therapeutic benefits, safety concerns around synthetic antioxidants, and accessibility and affordability [61, 62, 99].

Endogenous antioxidants	Description	References
Glutathione (GSH)	The most abundant intracellular non-protein thiol plays a central role in cellular detoxification and redox homeostasis.	[95]
Alpha-lipoic acid (ALA)	Naturally occurring antioxidants in plants and animals improve insulin sensitivity and glycemic control.	[95]
Coenzyme Q10 (CoQ10)	Antioxidants found in all body cells may improve insulin sensitivity and glycemic control.	[96]
Carotenoids	The pigment group found in fruits and vegetables has antioxidant and anti-inflammatory properties.	[96]
Vitamin C	Antioxidant found in citrus fruits and vegetables improves insulin sensitivity and glycemic control.	[96]
Vitamin E	Fat-soluble antioxidants in vegetable oils, nuts, seeds, and whole grains have antioxidant and anti-inflammatory properties.	[97]
Superoxide dismutase (SOD)	A main antioxidant enzyme that helps minimize oxidative stress by reducing the susceptibility of pancreatic islets to oxidative stress.	[97]
Glutathione peroxidase (GPX)	An antioxidant enzyme that helps protect against oxidative stress by reducing the levels of hydrogen peroxide and other free radicals.	[97]
Catalase	An antioxidant enzyme that helps protect against the expression of pro-inflammatory cytokines and can protect pancreatic islets against hydrogen peroxide and streptozotocin toxicity.	[97]
Peroxiredoxin (PRX)	Peroxiredoxin is an antioxidant enzyme that helps protect against oxidative stress by reducing hydrogen peroxide.	[98]

Table 1.
List of some endogenous antioxidants.

Supplementary antioxidants can effectively combat oxidative stress problems by scavenging free radicals. Commercially accessible antioxidant medications include multivitamins, multimineral supplements, pharmaceutical products with ascorbic acid, lipoic acid, sodium metabisulfite, propyl gallate, butylated hydroxytoluene, butylated hydroxyanisole, and cysteine. They are widely used in the food and pharmaceutical industries, although they have unfavorable side effects such as diarrhea and stomach discomfort [63, 64]. Additionally, reports suggest that clearing ROS too quickly could be harmful, impairing specific vital intracellular signaling and metabolic processes while making cells more vulnerable to infections [100].

Dietary antioxidants such as coffee, tea, red wine, honey, vitamins, and common Herbs and oils that show anti-diabetic effects also enhance the diabetic status by directing glucose metabolism, improving insulin secretion and insulin resistance, strengthening vascular functions, and managing the levels of HbA1c and oxidative stress markers [65]. Food and drinks contain various antioxidants, including polyphenols, flavonoids, bioflavonoids, and aromatic bio-phenols, which can lower oxidative stress and alter cellular defenses. Polyphenols are naturally occurring and artificially produced compounds with aromatic phenolic rings found in fruits, nuts, berries, green leafy vegetables, coffee, tea, red wine, olive oil, and honey [68]. Dietary antioxidants that contain polyphenols reduce the incidence of T2D in two ways: (1) through an insulin-dependent mechanism that protects and stimulates the growth of pancreatic islet β-cells while lowering oxidative stress and β-cell apoptosis, which turns on the insulin signaling and secretion; or (2) through an insulin-independent mechanism that reduces intestinal glucose absorption, inhibits the activity of digestive enzymes, and regulates the liver's glucose discharge [69, 71]. A list of some exogenous antioxidants is given in **Table 2**.

3.2.1 Viscosol and PTP1B inhibition

Numerous PTP1B inhibitors have been investigated, and the results have been encouraging in managing diabetes mellitus and its associated complications. PTP1B inhibitors can treat diseases related to insulin resistance by increasing the sensitivity of the insulin receptor [75].

The antidiabetic effect of the individual isolated polyphenolic component from aerial parts of a plant, *Dodonaea viscosa*, in the methanolic extract was investigated in a study by Ziauddin et al. *Dodonaea viscosa* is a major source of PTP inhibitors. The antidiabetic and hypoglycemic effect of *D. viscosa* extract was confirmed by phytochemical analysis, which isolated and identified PTP1B inhibitors. Compound 4 (viscosol), which is 5,7-dihydroxy-3,6-dimethoxy-2-(4- methoxy-3-(3-methyl but-2-enyl) phenyl)-4H-chromen-4-one, was discovered to be the most potent agent against PTP1B. The potent bioactive compound was isolated as a pale-yellow solid (molecular formula C23H24O7) with a molecular weight of 412.1522. Viscosol, which is a potent inhibitor of PTP1B, has an IC50 value of 13.5 μM and exhibits more fold inhibitory activity than other isolated compounds [103]. Viscosol is an inhibitor of protein tyrosine phosphatases (PTPs), such as PTP1B. The in-vivo analysis of viscosol has confirmed its antidiabetic effect. By inhibiting PTP1B, viscosol prevents the dephosphorylation of insulin receptors (IR) and insulin receptor substrates (IRS), thus allowing the PI3K/AKT signaling pathway to remain active [60].

Exogenous antioxidants	Description	References
L-Carnitine	An amino acid that helps protect against oxidative stress and inflammation.	[101]
Ruboxistaurin	It is a synthetic antioxidant that inhibits protein kinase C and has been shown to improve glycemic control.	[101]
LY 333531	It is a synthetic antioxidant that inhibits protein kinase C and has been shown to improve glycemic control.	[101]
Melatonin	A hormone that has antioxidant properties and has been shown to improve insulin sensitivity and glycemic control.	[101]
Polyphenols	A group of antioxidants in plant-based foods have anti-inflammatory and antioxidant properties.	[101]
Quercetin	A flavonoid that has antioxidant and anti-inflammatory properties and has been shown to improve insulin sensitivity and glycemic control.	[102]
Anthocyanin	A flavonoid that has antioxidant and anti-inflammatory properties and has been shown to improve insulin sensitivity and glycemic control.	[102]
Flavonoids	A group of antioxidants in plant-based foods have anti-inflammatory and antioxidant properties.	[102]
Carotenoids	A group of antioxidants in plant-based foods have antioxidant and anti-inflammatory properties.	[102]
Selenium	This mineral has antioxidant properties and has been shown to improve insulin sensitivity and glycemic control.	[102]
Curcumin	A polyphenol found in turmeric has antioxidant and anti-inflammatory properties and has been shown to improve insulin sensitivity and glycemic control.	[102]

Table 2.
List of some exogenous antioxidants and therapeutic compounds.

4. Vascular complications and oxidative stress in diabetes

4.1 Endothelial dysfunction and vascular complications

The vascular endothelium is a constant lining within the circulatory system (internal cavities, cardiac valves, and blood arteries) responsible for sustaining vascular tension, angiogenesis, and vascular hemostasis. It also offers essential surfaces for anti-inflammatory, antithrombotic, and antioxidant activities. Endothelial dysfunction can reduce vasodilation capability or, more broadly, any alterations that impact the function of endothelium homeostasis. The unique position of the endothelium allows it to detect changes in blood flow and cardiovascular pressure and respond by releasing vasoactive substances. This dynamic regulation is essential for maintaining vascular homeostasis, achieved through a delicate balance between the relaxing and contracting factors produced by the endothelium. When this balance is compromised, it puts the vasculature at risk for vasoconstriction, atherosclerosis, leukocyte adhesion, platelet activation, mitogenesis, pro-oxidation, thrombosis, poor coagulation, and vascular inflammation [76]. The current understanding of the relationship between endothelial dysfunction and the vascular complication of diabetes is incomplete. Patients with diabetes commonly develop severe diabetic vascular problems, which have a significant negative impact on their quality of life and prospects. Risk factors include high blood pressure, blood glucose, and cholesterol lipids. These vasodilation factors cause endothelial dysfunction in the vascular endothelium, which manifests as decreased NO biosynthesis or uncoupling of endothelial nitric oxide synthase (eNOS), increased inflammation and oxidative stress, endothelial-mesenchymal transition, senescence, and eventually cell death [77].

The primary cause of the health burden associated with diabetes is diabetic vascular complications. Chronic hyperglycemia elevates the risk of diabetic vascular complications, which include micro- and macrovascular complications. The microvascular complications consist of diabetic microangiopathy, diabetic nephropathy, diabetic retinopathy, and diabetic neuropathy, while the macrovascular complications comprise peripheral artery disease (PAD), cardiovascular diseases (CVD), and cerebrovascular disorders [78]. Under hyperglycemic circumstances, vascular endothelial dysfunction is a crucial mediator of microvascular and macrovascular problems. Intracellular hyperglycemia in diabetic microvasculature damages the vascular endothelium by activating four biochemical pathways: the hexosamine pathway, polyol pathway, protein kinase C (PKC) pathway, and advanced glycation end products (AGE) pathway. These pathways transfer glucose and its intermediates, producing excess reactive oxygen species. Inappropriate regulation of growth factors and cytokines, epigenetic modifications involving DNA modifications in response to intracellular changes, and variations in non-coding RNAs, particularly microRNAs, are also responsible for endothelial dysfunction (**Figure 3**) [79].

Endothelial dysfunction leads to the initiation and development of atherosclerosis. Endothelial dysfunction effectively contributes to the development of lesion formation by promoting both the early and late processes of atherosclerosis. These processes include adhesion molecule overexpression, increased leukocyte adhesion and

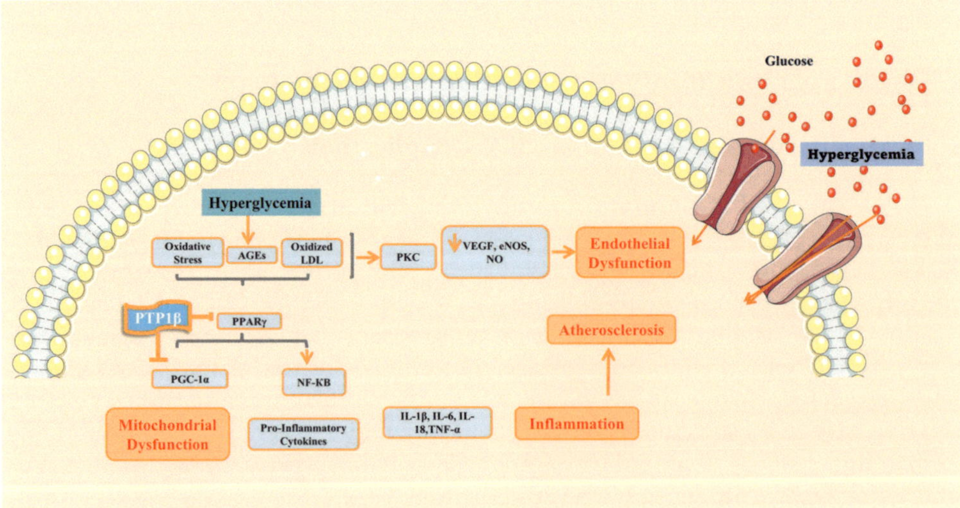

Figure 3.
The impact of hyperglycemia on endothelial dysfunction, inflammation, and mitochondrial dysfunction in type 2 diabetes mellitus, highlighting the role of PTP1B.

chemokine production, elevated cell permeability, accelerated LDL oxidation, platelet stimulation, cytokine synthesis, and proliferation and migration of vascular smooth muscle cells [80]. In coronary arteries with established atherosclerosis, impaired endothelium-dependent vasodilation leads to paradoxical vasoconstriction, which can cause myocardial ischemia and decreased myocardial perfusion. Furthermore, endothelial dysfunction predicts the lesion's susceptibility to rupture and actively modifies the architecture of the plaque. Unstable coronary syndromes may arise because of endothelial dysfunction in atherosclerotic arteries through this vasoconstrictor and inflammatory process [81].

4.2 Role of oxidative stress in atherosclerosis

Atherosclerosis, a continuous inflammatory disorder of the vascular system, is the primary cause of cardiac diseases worldwide. Oxidative stress, resulting from excessive production of ROS, is a significant risk factor for the onset and progression of atherosclerosis. ROS is crucial for regulating vascular health through their potent communication abilities but also triggers pro-atherogenic mechanisms like inflammation, impaired lipid metabolism, and endothelial dysfunction [82]. Vascular ROS are primarily produced by enzymes such as xanthine oxidase, uncoupled endothelial nitric oxide synthases (eNOS), NADPH oxidases, and mitochondrial electron transport chains. These enzymes transfer electrons from their specific substrates to oxygen molecules, catalyzing oxygen reduction. Both signaling and cellular homeostasis depend on basal levels of ROS, but oxidative stress can be driven by increased ROS formation or decreased antioxidant activity. The "kindling-bonfire radicals" hypothesis suggests that ROS sources can be divided into primary (mitochondrial respiratory chain and NADPH oxidases) and secondary (xanthine oxidase and uncoupled eNOS), with crosstalk between these groups activating further ROS production [104].

Superoxide (O_2^-) is created by various enzymatic sources, reacting with NO from eNOS and inducible NO synthase (iNOS) to form peroxynitrite radicals ($ONOO^-$). These radicals can scavenge tetrahydrobiopterin (BH4), causing eNOS uncoupling and producing more superoxide. SOD enzymes in the extracellular matrix and mitochondria convert superoxide ions into less harmful hydrogen peroxide (H_2O_2) molecules. NOX_4 and xanthine oxidase can also directly create hydrogen peroxide, which reacts with low-density lipoproteins (LDL) to produce oxidized LDL (ox-LDL). This crucial pathogenic phase in atherosclerosis promotes foam cell development, endothelial dysfunction, and inflammation [105]. OxLDL accumulates in artery walls prone to atherosclerosis and upregulates selectins P and E, VCAM-1, and other cell adhesion molecules on endothelial cells. These adhesion molecules recruit leukocytes, which invade the sub-endothelial region of the arterial wall, differentiating into macrophages and internalizing transformed lipoproteins to become foam cells, the initial sign of an atherosclerotic lesion [106].

5. Current research and advancements

At present, many kinds of medicines have been authorized for T2DM management. These include meglitinides, biguanides, thiazolidinediones (TZD), sulfonylurea, and inhibitors of α-glucosidase, SGLT2, and DPP-4. Additional authorized medications for the treatment of T2DM show various side effects that include a higher risk of cardiovascular diseases, hypoglycemia, infections in the urinary tract, and sudden body weight variations [92]. Despite the growing number of therapeutic choices in recent decades, several treatments have shown limits and disadvantages. Although usually prescribed medication for individuals with T2DM is metformin, we still lack proper information about its mode of action. Furthermore, its use is associated with gastrointestinal issues [93].

There are already many authorized medications for treating T2DM, but the number of filed clinical studies for finding any novel formulation has still exceeded 7000. Therefore, developing new treatment strategies to effectively cure T2DM and relevant macro- and microvascular problems is crucial. Innovative therapies are anticipated to provide additional benefits compared to the present treatments. These benefits include the possibility of weight reduction, no danger of hypoglycemia, enhanced drug delivery methods, and reduced frequency of usage [107].

A recent study compared the inhibitory activities of red wine (RW), red onion extract (RE), yellow wine (YW), and yellow onion extract (YE) against PTP1B, finding that all were effective. Still, YW was the most potent, exhibiting activity ten times stronger than ursolic acid (positive control). Water extracts were more effective than 70% ethanol extracts. In a HepG2 cellular model of insulin resistance, water extract reduced PTP1B expression in a concentration-dependent manner and increased insulin-induced glucose absorption. These findings suggest that onion extract contains compounds like quercetin that enhance insulin sensitivity. Given the lower toxicity of water extracts compared to ethanol extracts, likely due to highly glycosylated molecules increasing polarity and reducing toxicity, water extracts appear to be a superior option for treating T2DM. Identifying the specific elements in water extracts that improve glucose absorption by enhancing HepG2 cell responsiveness to insulin is valuable for future research [108, 109].

α-D-glucoside glucohydrolase, or simply α-glucosidase, is crucial for carbohydrate metabolism, converting disaccharides into monosaccharides and managing postprandial hyperglycemia [110]. Inhibitors of α-glucosidase can reduce postprandial hyperglycemia, making them beneficial for individuals with T2DM. Recent studies assessed the inhibitory activity of yellow wine (YW), red onion extract (RE), yellow onion extract (YE), and red wine (RW) on α-Glucosidase. YE exhibited the most potent inhibition, surpassing the positive control by a factor of 20, followed by YW, which was 12 times more powerful than acarbose. Managing postprandial hyperglycemia is crucial as it is more significantly associated with cardiovascular morbidity and mortality than fasting hyperglycemia, making YE and YW potentially advantageous for diabetes treatment [109]. Diabetes mellitus accelerates the formation of advanced glycation end-products (AGEs) due to increased ROS production, leading to various complications such as retinopathy, nephropathy, and other microvascular damage. Inhibitors that scavenge radicals or chelate metals can help manage high levels of AGEs. YE and RE extracts showed about five times the activity compared to aminoguanidine, while YW exhibited twice the inhibitory effect [109]. Ethanol extracts (70%) showed higher antioxidant activities than water extracts due to elevated levels of phenolic and flavonoid contents like quercetin. This suggests that 70% ethanol is a suitable choice for extracting natural antioxidants, with the peel of red onion outperforming yellow onion in terms of insulin sensitizers and PTP1B inhibitors [109].

Diabetic individuals often exhibit dyslipidemia, oxidative stress, insulin resistance, and heightened systemic inflammation, leading to complications like nephropathy, retinopathy, hypertension, and cardiovascular disease (CVD) [94]. Aerobic training (AT) provides benefits such as increased insulin receptor signaling, elevated GLUT-4 levels, more muscle capillaries and mitochondria, and improved glucose absorption, enhancing glycogen synthase and hexokinase activity. AT also helps regulate free fatty acids, reducing visceral fat and overall body weight [111]. Vitamin D increases insulin synthesis in the pancreas and GLUT-4 levels in skeletal muscle while inhibiting NF-Kb to reduce insulin resistance [112, 113]. Elevated irisin levels after exercise can reduce obesity and improve insulin resistance. Studies show that combining exercise with vitamin D supplementation significantly increases serum irisin, suggesting potential benefits for diabetes treatment by reducing PTP1B expression and enhancing insulin function [114–116]. Vitamin D and AT have been shown to markedly reduce PTP1B expression and significantly improve glycemia, suggesting a viable intervention in lowering insulin resistance and glycemia control in T2DM patients [117–119].

Some *in-vivo* antioxidant therapies are also on the focused in latest research having a sound impact. A recent study has examined the effects of cold atmospheric plasma (CAP) therapy on diabetes and its related complications, both *in vivo* and *in vitro*. In *in vitro* studies, the effects of CAP on glycated glutathione peroxidase (GPx) enzyme, an essential antioxidant enzyme, were investigated. The exposure of animals to CAP led to 30% elevation of glycated GPx activity and differences in protein structure, which was analyzed by circular dichroism and fluorescence studies [120]. Based on these *in vitro* findings, the effect of CAP treatment was evaluated in an animal model of diabetes. T2DM mice were topically exposed to CAP for 600 seconds on the skin, and then the levels of multiple biochemical factors were assayed 1 week after irradiation. Results showed that CAP treatment decreased blood glucose levels, diminished advanced glycation end products (AGEs), inflammatory cytokines, advanced oxidation protein products (AOPP), and oxidative stress markers including malondialdehyde (MDA) and elevated glutathione peroxidase function as antioxidant support. These results support the idea that CAP may serve as an adjuvant therapy for diabetes, in view of its beneficial effects on oxidative stress, inflammation, and AGE formation.

However, more clinical investigations are required to comprehensively assess this method's therapeutic capabilities because CAP can restore the equilibrium of redox reactions and alleviate the harmful consequences of high blood sugar levels [120].

A study published in January 2023 explored the capacity of antioxidant phytochemicals and polyphenols in mitigating and managing diabetes complications. Supported by several other *in vivo* studies, this research highlights the beneficial effects of these natural compounds. Polyphenols have demonstrated anti-diabetic, antioxidant, and anti-inflammatory characteristics that may aid in preventing diabetic nephropathy, neuropathy, and retinopathy. Phytochemicals derived from plants such as *Lycium barbarum* (wolfberry) have shown protective properties on spermatogenesis, erectile dysfunction, and other diabetes-related issues by enhancing antioxidant activity [121]. The review summarizes how different polyphenols, such as UCP2 and mitochondrial complex 1, might influence the body's antioxidant systems to protect against diabetes-related issues. The focus is on elucidating that oxidative stress induced by hyperglycemia plays a central role in the development of diabetes and further complications. Antioxidant phytochemicals and polyphenols can reduce oxidative damage by scavenging free radicals and reactive oxygen/nitrogen species, increasing the expression of naturally occurring antioxidant enzymes, suppressing the production of AGEs, and decreasing inflammation by regulating

Compound	Target	Mechanism	References
Filipino phytochemicals	PTP1B, DPP4 and SLGT2	Inhibition of targets.	[102]
Sulfonylureas & non-sulfonylureas	Insulin secretagogues	Stimulates insulin secretion from β-cells.	[123]
18α-glycyrrhetinic acid	11β-HSD1	Selective inhibition of 11β-HSD1.	[124]
Dehydrodieugenol B from *Ocimum tenuiflorum*	α-Amylase	Inhibition of human pancreatic α-amylase.	[124]
Amorfrutins	PPARγ	Activation of PPARγ transcriptional activity in adipocytes.	[124]
Curcumins from *Curcuma longa*	PTP1B, α-Glucosidase	Potential inhibition of PTP1B and α-Glucosidase.	[124]

Table 3.
Summary of recent studies on new targets and compounds against diabetes.

signal transmission within the body. The encouraging therapeutic potential of these natural antioxidant compounds in managing diabetes and its related problems suggests that they can restore redox equilibrium and reduce the harmful consequences of hyperglycemia [121].

A recent systematic review investigated the scientific evidence supporting quercetin, a flavonoid in many plant-based diets, as a treatment for diabetes and its associated complications. The researchers examined 34 papers that explored the impact of quercetin on diabetes complications in the laboratory and live organism environments. *In vitro* investigations revealed that quercetin suppressed glycation, inhibited the formation of AGEs, increased the activities of antioxidant enzymes such as GPx and SOD, and diminished oxidative stress by lowering levels of reactive oxygen species (ROS) and malondialdehyde (MDA) [122]. *In vivo* studies demonstrated that quercetin enhanced glucose absorption, reduced insulin resistance, lowered blood glucose levels, and safeguarded against diabetic complications such as nephropathy, neuropathy, and retinopathy in animal models. The promising data from both *in vitro* and *in vivo* studies support quercetin as a potential therapeutic agent for treating diabetes and its consequences. The antioxidant and anti-inflammatory properties of quercetin, along with its ability to block glycation and enhance insulin sensitivity, suggest that it is a valuable candidate for future research as a supplementary treatment for managing diabetes [122]. **Table 3** summarizes recent studies on novel targets and compounds to combat diabetes.

6. Conclusion

In summary, oxidative stress has been implicated in the origins and progress of T2DM. The overgeneration of ROS due to chronic hyperglycemia leads to insulin resistance and cellular injury. The most exciting option for therapy based on the findings so far may be antioxidant interventions, which can eliminate excess ROS, either endogenous or exogenous, respectively. Inhibition of PTP1B to improve insulin sensitivity is another potential strategy for curing T2DM. There will undoubtedly be future research to optimize antioxidant therapy or try novel inhibitors regarding antioxidants to develop efficient methods for managing diabetes.

Acknowledgements

I express my deepest gratitude to my supervisor, Dr. Imran Ullah Yousafzai, for his guidance and support.

Conflict of interest

The authors declare no conflict of interest.

Author details

Aamir Sohail, Muhammad Murtaza Hasnain, Muhammad Ehsan Ul Haq, Irum Nasir, Raza Sufyan, Mehroz Khan and Imran Ullah*
Quaid-i-Azam University, Islamabad, Pakistan

*Address all correspondence to: imranullah@qau.edu.pk

References

[1] Forouhi NG, Wareham NJ. Epidemiology of diabetes. Medicine. 2019;**47**(1):22-27

[2] Orban JC, Van Obberghen E, Ichai C. Acute complications of diabetes. In: Ichai C, Quintard H, Orban JC, editors. Metabolic Disorders and Critically Ill Patients. Cham: Springer; 2018. DOI: 10.1007/978-3-319-64010-5_15

[3] Eizirik D, Pasquali L, Cnop M. Pancreatic β-cells in type 1 and type 2 diabetes mellitus: Different pathways to failure. Nature Reviews. Endocrinology. 2020;**16**(7):349-362. Available from: https://www.nature.com/articles/s41574-020-0355-7

[4] Czech MP. Insulin action and resistance in obesity and type 2 diabetes. Nature Medicine. 2017;**23**(7):804-814

[5] Kopp W. How Western diet and lifestyle drive the pandemic of obesity and civilization diseases. Diabetes, Metabolic Syndrome and Obesity. 2019;**12**:2221-2236

[6] Nanda M, Sharma R, Mubarik S, Aashima A, Zhang K. Type-2 diabetes mellitus (T2DM): Spatial-temporal patterns of incidence, mortality and attributable risk factors from 1990 to 2019 among 21 world regions. Endocrine. 2022;**77**(3):444-454

[7] Zhu R, Zhou S, Xia L, Bao X. Incidence, morbidity and years lived with disability due to type 2 diabetes mellitus in 204 countries and territories: Trends from 1990 to 2019. Frontiers in Endocrinology. 2022;**13**:905538

[8] Liu J, Ren ZH, Qiang H, Wu J, Shen M, Zhang L, et al. Trends in the incidence of diabetes mellitus: Results from the global burden of disease study 2017 and implications for diabetes mellitus prevention. BMC Public Health. 2020;**20**:1-2

[9] Ye J, Wu Y, Yang S, Zhu D, Chen F, Chen J, et al. The global, regional and national burden of type 2 diabetes mellitus in the past, present and future: A systematic analysis of the global burden of disease study 2019. Frontiers in Endocrinology. 2023;**14**:1192629

[10] Azeem S, Khan U, Liaquat A. The increasing rate of diabetes in Pakistan: A silent killer. Annals of Medicine and Surgery (London). 2022;**79**:103901. Available from: https://www.ncbi.nlm.nih.gov/pmc/articles/PMC9289249/

[11] Pathak J, Rajneesh, Ahmed H, Singh DK, Singh PR, Kumar D, et al. Oxidative stress and antioxidant defense in plants exposed to ultraviolet radiation. In: Reactive Oxygen, Nitrogen and Sulfur Species in Plants: Production, Metabolism, Signaling and Defense Mechanisms. Hoboken, New Jersey, USA: Wiley; 2019. pp. 371-420

[12] Ogunro OB, Fakayode AE, Batiha GE. Involvement of antioxidants in the prevention of cellular damage. In: Importance of Oxidative Stress and Antioxidant System in Health and Disease. London, UK: IntechOpen; 2022

[13] Leyane TS, Jere SW, Houreld NN. Oxidative stress in ageing and chronic degenerative pathologies: Molecular mechanisms involved in counteracting oxidative stress and chronic inflammation. International Journal of Molecular Sciences. 2022;**23**(13):7273

[14] Ogunro OB, Fakayode AE, Batiha GES. Involvement of Antioxidants

in the Prevention of Cellular Damage. London, UK: IntechOpen. [Internet]. 2022. Available from: https://www.intechopen.com/chapters/84852

[15] Panigrahy SK, Bhatt R, Kumar A. Reactive oxygen species: Sources, consequences and targeted therapy in type 2 diabetes. Journal of Drug Targeting. 2017;**25**(2):93-101

[16] Prasun P. Role of mitochondria in the pathogenesis of type 2 diabetes mellitus. Journal of Diabetes and Metabolic Disorders. 2020;**19**(2):2017-2022

[17] González P, Lozano P, Ros G, Solano F. Hyperglycemia and oxidative stress: An integral, updated and critical overview of their metabolic interconnections. International Journal of Molecular Sciences. 2023;**24**(11):9352

[18] De Gaetano A, Gibellini L, Zanini G, Nasi M, Cossarizza A, Pinti M. Mitophagy and oxidative stress: The role of aging. Antioxidants. 2021;**10**(5):794

[19] Saltiel AR. Insulin signaling in health and disease. The Journal of Clinical Investigation. 2021;**131**(1):e142241

[20] Huang X, Liu G, Guo J, Su Z. The PI3K/AKT pathway in obesity and type 2 diabetes. International Journal of Biological Sciences. 2018;**14**(11):1483

[21] Singh S, Grewal AS, Grover R, Sharma N, Chopra B, Dhingra AK, et al. Recent updates on development of protein-tyrosine phosphatase 1B inhibitors for treatment of diabetes, obesity and related disorders. Bioorganic Chemistry. 2022;**121**:105626

[22] Coronell-Tovar A, Pardo JP, Rodríguez-Romero A, Sosa-Peinado A, Vásquez-Bochm L, Cano-Sánchez P, et al. Protein tyrosine phosphatase 1B (PTP1B) function, structure, and inhibition strategies to develop antidiabetic drugs. FEBS Letters. 2024;**598**(15):1811-1838

[23] Lorenzo M, Fernández-Veledo S, Vila-Bedmar R, Garcia-Guerra L, De Alvaro C, Nieto-Vazquez I. Insulin resistance induced by tumor necrosis factor-α in myocytes and brown adipocytes. Journal of Animal Science. 2008;**86**(Suppl. 14):E94-E104. DOI: 10.2527/jas.2007-0462

[24] Yang L, Sun YY, Liu YR, Yin NN, Bu FT, Yu HX, et al. PTP1B promotes macrophage activation by regulating the NF-κB pathway in alcoholic liver injury. Toxicology Letters. 2020;**319**:11-21

[25] Teimouri M, Hosseini H, Arab-Sadeghabadi Z, Babaei-Khorzoughi R, Gorgani-Firuzjaee S, Meshkani R. The role of protein tyrosine phosphatase 1B (PTP1B) in the pathogenesis of type 2 diabetes mellitus and its complications. Journal of Physiology and Biochemistry. 2022;**78**(2):307-322

[26] Sharma B, Xie L, Yang F, Wang W. Recent advances in PTP1B inhibitors and their biomedical applications. European Journal of Medicinal Chemistry. 2020;**203**:112658. Available from: https://www.sciencedirect.com/science/article/pii/S0223523420303469

[27] Makhijani P, Basso PJ, Chan YT, Chen N, Baechle J, Khan S, et al. Regulation of the immune system by the insulin receptor in health and disease. Frontiers in Endocrinology (Lausanne). 2023;**14**:1092386

[28] Chandrasekaran P, Weiskirchen R. Cellular and molecular mechanisms of insulin resistance. Current Tissue Microenvironment Reports. 2024;**5**(2):1-12

[29] White MF. Mechanism of insulin action. In: Textbook of Diabetes. Hoboken, New Jersey, USA: Wiley; 2017. pp. 114-132

[30] Taheri R, Mokhtari Y, Yousefi AM, Bashash D. The PI3K/Akt signaling axis and type 2 diabetes mellitus (T2DM): From mechanistic insights into possible therapeutic targets. Cell Biology International. 2024;**48**(8):1049-1068

[31] Lieu CV, Loganathan N, Belsham DD. Mechanisms driving palmitate-mediated neuronal dysregulation in the hypothalamus. Cells. 2021;**10**(11):3120

[32] Kunter I, Zabib N. A review of oxidants and antioxidants in biological systems. Journal of Pharmaceutical Sciences. [Internet]. 2020;**3**(1):64-72. Available from: http://i-rep.emu.edu.tr:8080/xmlui/bitstream/handle/11129/4975/EMU%20Journal%20of%20Pharmaceutical%20Sciences%20-%20%20Vol%203%20No%201%202020.pdf?sequence=1#page=74

[33] Verma S, Sharma S. Protein tyrosine phosphatase as potential therapeutic target in various disorders. Current Molecular Pharmacology. 2018;**11**(3):191-202

[34] Sisein EA. Biochemistry of free radicals and antioxidants. Scholars Academic Journal of Biosciences. 2014;**2**(2):110-118

[35] Dziąbowska-Grabias K, Sztanke M, Zając P, Celejewski M, Kurek K, Szkutnicki S, et al. Antioxidant therapy in inflammatory bowel diseases. Antioxidants (Basel). 2021;**10**(3):412. Available from: https://www.mdpi.com/2076-3921/10/3/412/htm

[36] Mirończuk-Chodakowska I, Witkowska AM, Zujko ME. Endogenous non-enzymatic antioxidants in the human body. Advances in Medical Sciences. 2018;**63**(1):68-78

[37] Ungaro R, Chang HL, Cote-Daigneaut J, Mehandru S, Atreja A, Colombel JF. Statins associated with decreased risk of new onset inflammatory bowel disease. The American Journal of Gastroenterology. 2016;**111**(10):1416-1423

[38] Bratovcic A. Antioxidant enzymes and their role in preventing cell damage. Acta Scientific Nutritional Health. 2020;**4**:1-7

[39] Moura FA, Goulart MO, Campos SB, da Paz Martins AS. The close interplay of nitro-oxidative stress, advanced glycation end products and inflammation in inflammatory bowel diseases. Current Medicinal Chemistry. 2020;**27**(13):2059-2076

[40] Bezzio C, Festa S, Saibeni S, Papi C. Chemoprevention of colorectal cancer in ulcerative colitis: Digging deep in current evidence. Expert Review of Gastroenterology & Hepatology. 2017;**11**(4):339-347

[41] Mak JW, So J, Tang W, Yip TC, Leung WK, Li M, et al. Cancer risk and chemoprevention in Chinese inflammatory bowel disease patients: A population-based cohort study. Scandinavian Journal of Gastroenterology. 2020;**55**(3):279-286

[42] Dziabowska-Grabias K, Sztanke M, Zajac P, Celejewski M, Kurek K, Szkutnicki S, et al. Antioxidant therapy in inflammatory bowel diseases. Antioxidants. 2021;**10**:412

[43] Bryniarski P, Nazimek K, Marcinkiewicz J. Immunomodulatory activity of the most commonly used antihypertensive drugs—Angiotensin converting enzyme inhibitors and

angiotensin II receptor blockers. International Journal of Molecular Sciences. 2022;**23**(3):1772

[44] Tian T, Wang Z, Zhang J. Pathomechanisms of oxidative stress in inflammatory bowel disease and potential antioxidant therapies. Oxidative Medicine and Cellular Longevity. 2017;**2017**(1):4535194

[45] Irrazabal T, Thakur BK, Croitoru K, Martin A. Preventing colitis-associated colon cancer with antioxidants: A systematic review. Cellular and Molecular Gastroenterology and Hepatology. 2021;**11**(4):1177-1197

[46] Wang W, Kang PM. Oxidative stress and antioxidant treatments in cardiovascular diseases. Antioxidants (Basel). 2020;**9**(12):1292. Available from: https://www.mdpi.com/2076-3921/9/12/1292/htm

[47] Gagnière J, Bonnet M. Molecular mechanism underlying the actions of antioxidant molecules in digestive disorders. In: Gastrointestinal Tissue. Cambridge, Massachusetts, USA: Academic Press; 2017. pp. 197-216

[48] Sen S, Chakraborty R. Herbs, gastrointestinal protection, and oxidative stress. In: Gastrointestinal Tissue. Cambridge, Massachusetts, USA: Academic Press; 2017. pp. 259-274

[49] Xiao HT, Wen B, Shen XC, Bian ZX. The potential of plant-sourced phenols for inflammatory bowel disease. Current Medicinal Chemistry. 2018;**25**(38):5191-5217

[50] Orellana-Urzúa S, Briones-Valdivieso C, Chichiarelli S, Saso L, Rodrigo R. Potential role of natural antioxidants in countering reperfusion injury in acute myocardial infarction and ischemic stroke. Antioxidants. 2023;**12**(9):1760

[51] Pereira CP, Souza AC, Vasconcelos AR, Prado PS. Antioxidant and anti-inflammatory mechanisms of action of astaxanthin in cardiovascular diseases. International Journal of Molecular Medicine. 2021;**47**(1):37-48

[52] Gao HL, Yu XJ, Zhang Y, Wang CL, Lei YM, Yu JY, et al. Astaxanthin ameliorates blood pressure in salt-induced prehypertensive rats through ROS/MAPK/NF-κ B pathways in the hypothalamic paraventricular nucleus. Cardiovascular Toxicology. 2021;**21**:1045-1057

[53] Roth Flach RJ, Su C, Bollinger E, Cortes C, Robertson AW, Opsahl AC, et al. Myeloperoxidase inhibition in mice alters atherosclerotic lesion composition. PLoS One. 2019;**14**(3):e0214150

[54] Loyer X, Potteaux S, Vion A-C, Guérin CL, Boulkroun S, Rautou P-E, et al. Inhibition of microRNA-92a prevents endothelial dysfunction and atherosclerosis in mice. Circulation Research. 2014;**114**(3):434-443

[55] Chaudhuri RD, Rana S, Datta K, Sarkar S. Key cellular effectors in ROS-mediated cardiac diseases. In: Modulation of Oxidative Stress in Heart Disease. Singapore: Springer; 2019. pp. 151-195

[56] Yang J, Zhang R, Zhao H, Qi H, Li J, Li JF, et al. Bioinspired copper single-atom enzyme as a superoxide dismutase-like antioxidant for sepsis treatment. Exploration. 2022;**2**(4):20210267. Available from: https://onlinelibrary.wiley.com/doi/full/10.1002/EXP.20210267

[57] Najam A, Ahmad S, Abid R, Ali H, Husnain M, Aziz T, et al. Immune-adjuvant effect of vitamin A and

probiotics supplementation on humoral response to cell culture rabies vaccine in rabbits. 3 Biotech. 2023;**13**(7):290. Available from: https://pubmed.ncbi.nlm.nih.gov/37323857/

[58] Javid AZ, Hosseini SA, Gholinezhad H, Moradi L, Haghighi-Zadeh MH, Bazyar H. Antioxidant and anti-inflammatory properties of melatonin in patients with type 2 diabetes mellitus with periodontal disease under non-surgical periodontal therapy: A double-blind, placebo-controlled trial. Diabetes, Metabolic Syndrome and Obesity. 2020;**13**:753-761. Available from: https://www.tandfonline.com/action/journalInformation?journalCode=dmso20

[59] Fatima MT, Bhat AA, Nisar S, Fakhro KA, Akil AS. The role of dietary antioxidants in type 2 diabetes and neurodegenerative disorders: An assessment of the benefit profile. Heliyon. 2023;**9**(1):e12698

[60] Sohail A, Fayyaz H, Muneer H, Raza I, Ikram M, Uddin Z, et al. Targeted inhibition of protein tyrosine phosphatase 1b by viscosol ameliorates type 2 diabetes pathophysiology and histology in diabetic mouse model. BioMed Research International. 2022;**2022**(1):2323078

[61] Hassan W, Noreen H, Rehman S, Kamal MA, da Rocha JB. Association of oxidative stress with neurological disorders. Current Neuropharmacology. 2022;**20**(6):1046

[62] Saha S, Sadhukhan P, Sil PC. Beneficial upshots of naturally occurring antioxidant compounds against neurological disorders. In: Neuroprotective Natural Products: Clinical Aspects and Mode of Action. Hoboken, New Jersey, USA: Wiley; 2017. pp. 23-56

[63] Yaribeygi H, Sathyapalan T, Sahebkar A. Effects of novel antidiabetes agents on apoptotic processes in diabetes and malignancy: Implications for lowering tissue damage. Life Sciences. 2019;**231**:116542. Available from: https://www.sciencedirect.com/science/article/pii/S0024320519304588

[64] Khandelwal B, Gupta C, Thomas L, Miraj SS, Rao M, Dey R, et al. Free radical biology of diabetes mellitus. In: Free Radical Biology of & Endocrine, Metabolic Immune Disorders. Sharjah, UAE: Bentham Science; 2022. p. 1

[65] Halim M, Halim A. The effects of inflammation, aging and oxidative stress on the pathogenesis of diabetes mellitus (type 2 diabetes). Diabetes & Metabolic Syndrome: Clinical Research & Reviews. 2019;**13**(2):1165-1172

[66] Ahmad S, Khan H, Shahab U, Rehman S, Rafi Z, Khan MY, et al. Protein oxidation: An overview of metabolism of sulphur containing amino acid, cysteine. Frontiers in Bioscience (Scholar Edition). 2017;**9**(1):71-87

[67] Rezayian M, Niknam V, Ebrahimzadeh H. Oxidative damage and antioxidative systems in algae. Toxicology Reports. 2019;**6**:1309-1313

[68] Foret MK, Lincoln R, Do Carmo S, Cuello AC, Cosa G. Connecting the "dots": From free radical lipid autoxidation to cell pathology and disease. Chemical Reviews. 2020;**120**(23):12757-12787

[69] Jacob C, Fujii J, Homma T, Osaki T. Superoxide radicals in the execution of cell death. Antioxidants (Basel). 2022;**11**(3):501. Available from: https://www.mdpi.com/2076-3921/11/3/501

[70] Di Mascio P, Martinez GR, Miyamoto S, Ronsein GE, Medeiros MH,

Cadet J. Singlet molecular oxygen reactions with nucleic acids, lipids, and proteins. Chemical Reviews. 2019;**119**(3):2043-2086

[71] Drabińska N. Current perspective about the effect of a ketogenic diet on oxidative stress–a review. Polish Journal of Food and Nutrition Sciences. 2024;**74**(1):92-105

[72] Poetsch AR. The genomics of oxidative DNA damage, repair, and resulting mutagenesis. Computational and Structural Biotechnology Journal. 2020;**18**:207-219

[73] Karimi M, Roudini K, Zendehdel A, Toorang F, Ebrahimpour-Koujan S. DNA oxidation-based analysis: A new approach to assessing the relationship between nutrition and cancer. Basic & Clinical Cancer Research. 2023;**13**(4):258-271

[74] Cao R, Tian H, Zhang Y, Liu G, Xu H, Rao G, et al. Signaling pathways and intervention for therapy of type 2 diabetes mellitus. Molecular and Clinical Oncology. 2023;**4**(3):100178. Available from: https://onlinelibrary.wiley.com/doi/abs/10.1002/mco2.283

[75] Bansal S, Mahendiratta S, Agrawal M, Kumar S, Sharma AR, Garg N, et al. Role of protein tyrosine phosphatase 1B inhibitor in central insulin resistance and associated cognitive deficits. Brain Research Bulletin. 2021;**171**:113-125

[76] Mathur A, Pandey V, Kabra P. PHLPP: A putative cellular target during insulin resistance and type 2 diabetes. The Journal of Endocrinology. 2017;**233**(3):R185-R198. Available from: https://joe.bioscientifica.com/view/journals/joe/233/3/R185.xml

[77] Yung JHM, Giacca A. Role of c-Jun N-terminal kinase (JNK) in obesity and type 2 diabetes. Cells. 2020;**9**(3):706. Available from: https://www.ncbi.nlm.nih.gov/pmc/articles/PMC7140703/

[78] Wei H, Cui D. Pyroptosis and insulin resistance in metabolic organs. International Journal of Molecular Sciences. 2022;**23**(19):11638

[79] Feng J, Lu S, Ou B, Liu Q, Dai J, Ji C, et al. The role of JNk signaling pathway in obesity-driven insulin resistance. Diabetes, Metabolic Syndrome and Obesity. 2020;**13**:1399-1406

[80] Ighodaro OM. Molecular pathways associated with oxidative stress in diabetes mellitus. Biomedicine & Pharmacotherapy. 2018;**108**:656-662

[81] Oguntibeju OO. Type 2 diabetes mellitus, oxidative stress and inflammation: Examining the links. International Journal of Physiology, Pathophysiology and Pharmacology. 2019;**11**(3):45

[82] Fernandes-da-Silva A, Miranda CS, Santana-Oliveira DA, Oliveira-Cordeiro B, Rangel-Azevedo C, Silva-Veiga FM, et al. Endoplasmic reticulum stress as the basis of obesity and metabolic diseases: Focus on adipose tissue, liver, and pancreas. European Journal of Nutrition. 2021;**60**(6):2949-2960

[83] Mustapha S, Mohammed M, Azemi AK, Jatau AI, Shehu A, Mustapha L, et al. Status of endoplasmic reticulum stress in type II diabetes. Molecules. 2021;**26**(14):4362

[84] Villamar-Cruz O, Loza-Mejía MA, Arias-Romero LE, Camacho-Arroyo I. Recent advances in PTP1B signaling in metabolism and cancer. Bioscience Reports. 2021;**41**(11):BSR20211994

[85] Brautigan DL, Shenolikar S. Protein serine/threonine phosphatases: Keys

to unlocking regulators and substrates. Annual Review of Biochemistry. 2018;**87**:921-964

[86] Haeusler RA, McGraw TE, Accili D. Biochemical and cellular properties of insulin receptor signalling. Nature Reviews Molecular Cell Biology. 2018;**19**(1):31-44

[87] Salazar J, Chávez-Castillo M, Rojas J, Ortega A, Nava M, Pérez J, et al. Is "leptin resistance" another key resistance to manage type 2 diabetes? Current Diabetes Reviews. 2020;**16**(7):733-749

[88] Abdelsalam SS, Korashy HM, Zeidan A, Agouni A. The role of protein tyrosine phosphatase (PTP)-1B in cardiovascular disease and its interplay with insulin resistance. Biomolecules. 2019;**9**(7):286

[89] Shazmeen HIU, Rajoka MS, Asim Shabbir M, Umair M, llah IU, Manzoor MF, Nemat A, Abid M, Khan MR, Aadil RM. Role of stilbenes against insulin resistance: A review. Food Science & Nutrition. 2021;**9**(11):6389-6405

[90] Khaledi K, Hoseini R, Gharzi A. The impact of vitamin D on type 2 diabetes management: Boosting PTP1B gene expression and physical activity benefits in rats. Genes & Nutrition. 2024;**19**(1):1-1

[91] Yashin A, Yashin Y, Xia X, Nemzer B. Antioxidant activity of spices and their impact on human health: A review. Antioxidants (Basel). 2017;**6**(3):70

[92] Chaudhary P, Janmeda P, Docea AO, Yeskaliyeva B, Abdull Razis AF, Modu B, et al. Oxidative stress, free radicals and antioxidants: Potential crosstalk in the pathophysiology of human diseases. Frontiers in Chemistry. 2023;**11**:1158198

[93] Aziz MA, Diab AS, Mohammed AA. Antioxidant categories and mode of action. Antioxidants. 2019;**2019**:3-22

[94] Kontoghiorghes GJ, Kontoghiorghe CN. Prospects for the introduction of targeted antioxidant drugs for the prevention and treatment of diseases related to free radical pathology. Expert Opinion on Investigational Drugs. 2019;**28**(7):593-603

[95] Pruteanu LL, Bailey DS, Grădinaru AC, Jäntschi L. The biochemistry and effectiveness of antioxidants in food, fruits, and marine algae. Antioxidants. 2023;**12**(4):860

[96] Akbari A, Jelodar G, Nazifi S, Sajedianfard J. An overview of the characteristics and function of vitamin C in various tissues: Relying on its antioxidant function. Zahedan Journal of Research in Medical Sciences. 2016;**18**(11):e4037

[97] Didier AJ, Stiene J, Fang L, Watkins D, Dworkin LD, Creeden JF. Antioxidant and anti-tumor effects of dietary vitamins A, C, and E. Antioxidants. [Internet]. 2023;**12, 3**:632. Available from: https://www.mdpi.com/2076-3921/12/3/632

[98] Hewitt OH, Degnan SM. Antioxidant enzymes that target hydrogen peroxide are conserved across the animal kingdom, from sponges to mammals. Scientific Reports. 2023;**13**(1):2510

[99] Stoia M, Oancea S. Low-molecular-weight synthetic antioxidants: Classification, pharmacological profile, effectiveness and trends. Antioxidants. 2022;**11**(4):638

[100] Liguori I, Russo G, Curcio F, Bulli G, Aran L, Della-Morte D, et al. Oxidative stress, aging, and diseases.

Clinical Interventions in Aging. 2018;**13**:757-772. Available from: https://www.tandfonline.com/doi/abs/10.2147/CIA.S158513

[101] Arabshomali A, Bazzazzadehgan S, Mahdi F, Shariat-Madar Z. Potential benefits of antioxidant phytochemicals in type 2 diabetes. Molecules (Basel, Switzerland). [Internet]. 2023;**28**(20):7209. Available from: https://pubmed.ncbi.nlm.nih.gov/37894687/ [Accessed: December 1, 2023]

[102] Marrazzo G, Barbagallo I, Galvano F, Malaguarnera M, Gazzolo D, Frigiola A, et al. Role of dietary and endogenous antioxidants in diabetes. Critical Reviews in Food Science and Nutrition. 2014;**54**(12):1599-1616

[103] Uddin Z, Song YH, Ullah M, Li Z, Kim JY, Park KH. Isolation and characterization of protein tyrosine phosphatase 1B (PTP1B) inhibitory polyphenolic compounds from *Dodonaea viscosa* and their kinetic analysis. Frontiers in Chemistry. 2018;**6**:40. Available from: https://www.frontiersin.org/articles/10.3389/fchem.2018.00040

[104] Mustapha S, Mohammed M, Azemi AK, Jatau AI, Shehu A, Mustapha L, et al. Current status of endoplasmic reticulum stress in type II diabetes. Molecules. 2021;**26**(14):4362

[105] Tejero J, Shiva S, Gladwin MT. Sources of vascular nitric oxide and reactive oxygen species and their regulation. Physiological Reviews. 2019;**99**(1):311-379

[106] Sagris M, Theofilis P, Antonopoulos AS, Oikonomou E, Simantiris S, Papaioannou S, et al. Adhesion molecules as prognostic biomarkers in coronary artery disease. Current Topics in Medicinal Chemistry. 2023;**23**(6):481-490

[107] Perreault L, Skyler JS, Rosenstock J. Novel therapies with precision mechanisms for type 2 diabetes mellitus. Nature Reviews Endocrinology. 2021;**17**(6):364-377

[108] Watanabe J, Kawabata J, Kurihara H, Niki R. Isolation and identification of α-glucosidase inhibitors from tochu-cha (*Eucommia ulmoides*). Bioscience, Biotechnology, and Biochemistry. 1997, 1997;**61**:177-178. DOI: 10.1271/bbb.61.177

[109] Yang SJ, Paudel P, Shrestha S, Seong SH, Jung HA, Choi JS. In vitro protein tyrosine phosphatase 1B inhibition and antioxidant property of different onion peel cultivars: A comparative study. Food Science & Nutrition. 2018;**7**(1):205-215

[110] Kim YM, Jeong YK, Wang MH, Lee WY, Rhee HI. Inhibitory effect of pine extract on α-glucosidase activity and postprandial hyperglycemia. Nutrition. 2005;**21**:756-761. DOI: 10.1016/j.nut.2004.10.014

[111] Adderley NJ, Subramanian A, Toulis K, Gokhale K, Taverner T, Hanif W, et al. Obstructive sleep apnea, a risk factor for cardiovascular and microvascular disease in patients with type 2 diabetes: Findings from a population-based cohort study. Diabetes Care. 2020;**43**(8):1868-1877

[112] Climie RE, van Sloten TT, Bruno R-M, Taddei S, Empana J-P, Stehouwer CD, et al. Macrovasculature and microvasculature at the crossroads between type 2 diabetes mellitus and hypertension. Hypertension. 2019;**73**(6):1138-1149

[113] Bruce CR, Hawley JA. Improvements in insulin resistance with aerobic exercise training: A lipocentric

approach. Medicine and Science in Sports and Exercise. 2004;**36**(7):1196-1201

[114] Wang Y, Zhu J, DeLuca HF. Where is the vitamin D receptor? Archives of Biochemistry and Biophysics. 2012;**523**(1):123-133

[115] Foroughi M, Maghsoudi Z, Khayyatzadeh S, Ghiasvand R, Askari G, Iraj B. Relationship between non-alcoholic fatty liver disease and inflammation in patients with non-alcoholic fatty liver. Advanced Biomedical Research. 2016;**5**:28

[116] Hisamatsu T, Miura K, Arima H, Fujiyoshi A, Kadota A, Kadowaki S, et al. Relationship of serum irisin levels to prevalence and progression of coronary artery calcification: A prospective, population-based study. International Journal of Cardiology. 2018;**267**:177-182

[117] Giri D, Pintus D, Burnside G, Ghatak A, Mehta F, Paul P, et al. Treating vitamin D deficiency in children with type I diabetes could improve their glycaemic control. BMC Research Notes. 2017;**10**(1):1-5

[118] Polyzos SA, Anastasilakis AD, Efstathiadou ZA, Makras P, Perakakis N, Kountouras J, et al. Irisin in metabolic diseases. Endocrine. 2018;**59**:260-274

[119] Manna P, Achari AE, Jain SK. Vitamin D supplementation inhibits oxidative stress and upregulates SIRT1/AMPK/GLUT4 cascade in high glucose-treated 3T3L1 adipocytes and in adipose tissue of high fat diet-fed diabetic mice. Archives of Biochemistry and Biophysics. 2017;**615**:22-34

[120] Rezaeinezhad A, Eslami P, Mirmiranpour H, Ghomi H. The effect of cold atmospheric plasma on diabetes-induced enzyme glycation, oxidative stress, and inflammation; *in vitro* and *in vivo*. Scientific Reports. 2019;**9**(1):19958

[121] Akpoveso OOP, Ubah EE, Obasanmi G. Antioxidant phytochemicals as potential therapy for diabetic complications. Antioxidants. 2023;**12**(1):123

[122] Shi GJ, Li Y, Cao QH, Wu HX, Tang XY, Gao XH, et al. *In vitro* and *in vivo* evidence that quercetin protects against diabetes and its complications: A systematic review of the literature. Biomedicine & Pharmacotherapy. 2019;**109**:1085-1099

[123] Su J, Luo Y, Hu S, Lü T, Ouyang S. Advances in research on type 2 diabetes mellitus targets and therapeutic agents. International Journal of Molecular Sciences. 2023;**24**(17):13381-13381

[124] Vesa CM, Zaha D, editors. Research Progress of Plant Compounds for Diabetes and its Complications. [Internet]. MDPI - Multidisciplinary Digital Publishing Institute. Basel, Switzerland: MDPI; 2024. Available from: www.mdpi.com

Chapter 9

Role of Reactive Oxygen Species in Carcinogenesis and Polyphenols as an Emerging Therapeutic Intervention

Aparajita Das and Sarbani Giri

Abstract

Reactive oxygen species (ROS) are generated in the body as a by-product of cellular enzymatic reactions. Under normal conditions, an antioxidant defense mechanism in the body regulates the level of ROS produced and maintains a redox balance. However, in cases of metabolic disorder, chronic inflammation, or prolonged exposure to xenobiotics and environmental stressors, this balance is disturbed and leads to the generation of oxidative stress. ROS can attack the structural integrity of the major macromolecules of the body such as nucleic acids, lipids, and proteins leading to the generation of pathologies including cancer. Polyphenols have emerged as potent nutraceuticals that can not only augment the body's antioxidant defense system to combat the generated oxidative stress but can also selectively act as pro-oxidants in cancer cells, a dichotomous phenomenon that is being actively studied for implementation in cancer therapeutics. This chapter will present in a comprehensive manner the role of ROS in the pathogenesis of cancer and the application of pro-oxidant nature of polyphenols as chemotherapeutics.

Keywords: reactive oxygen species, cancer, polyphenols, pro-oxidant, cancer therapeutics

1. Introduction

Free radicals are molecules with an unpaired electron in their electronic configuration. As a consequence, they act as strong oxidizing agents and thereby react with or attack the major biomolecules of the body and take up electrons from them. The molecule, thus robbed of an electron, itself gets converted to a free radical which either initiates a chain reaction of oxidation-reduction or gets converted to another form with altered functionality and characteristics [1]. Oxygen-based free radicals and reactive non-radicals comprise reactive oxygen species (ROS) and have been listed in **Table 1**. Moderate amount of ROS is necessary for the maintenance of normal physiological processes like cellular proliferation and programmed cell death, redox signaling, wound healing, male fertility,

ROS species	Physiological sources
Superoxide anion ($O_2^{\cdot-}$)	Reduction of molecular oxygen. Spontaneous (leakage of an electron in electron transport chain or in the presence of reduced metal ion) or via enzymes
Perhydroxyl radical ($HOO^{\cdot}$)	Cytosolic protonation of $O_2^{\cdot-}$
Hydrogen peroxide (H_2O_2)	Dismutation of superoxide anions (spontaneous or enzyme-mediated) or double reduction of molecular oxygen by enzymes
Hydroxyl radical ($OH^{\cdot}$)	Reduction of H_2O_2 in the presence of Fe^{2+}/ Cu^{+} (Fenton reaction) or in the presence of $O_2^{\cdot-}$ (Haber-Weiss reaction)
Hypochlorous acid (HOCl)	Forms when H_2O_2 reacts with Cl^- in the presence of enzyme myeloperoxidase
Singlet oxygen ($^1O_2$)	Forms when H_2O_2 reacts with HOCl or with $O_2^{\cdot-}$. Also released as by-products of biochemical reactions involving primary or secondary reactive species
Ozone (O_3)	Oxidation of water or deoxidation of oxidized biomolecules by $^1O_2$
Alkoxyl radical, peroxyl radical, organic peroxide	Formed when primary reactive species react with cellular macromolecules/organic compounds

Source: [1–5].

Table 1.
Major reactive oxygen species.

immune responses, and aging to name a few [6, 7]. Antioxidants are compounds that keep in check the ROS content of the body either by directly quenching ROS by supplying electrons or indirectly by mediating reactions that facilitate reduction [8]. Endogenous antioxidants (enzymatic and non-enzymatic) are present inside the cell and also circulate in plasma. Edibles like diet and supplements can also provide antioxidants and are categorized as exogenous sources [8]. However, in case the level of ROS increases in cells in the presence of biotic or abiotic stimulants or in case of any pathological condition, an excess amount of ROS can harm the major biomolecules of the cell which leads to the development of several pathological conditions. Further increase in ROS level beyond a particular threshold induces activation of signaling pathways leading to cellular death such as apoptosis, autophagy, ferroptosis, and pyroptosis [9, 10]. In cancer therapeutics, one of the strategies to combat cancer is based on this phenomenon and involves the induction of cancer cell death by increasing the concentration of ROS in them.

Polyphenols are a group of phytochemicals, well known for their health advantages in humans [11]. The commercialization of polyphenol-based products and the application of polyphenols in various ailments is based on their potency as antioxidants. However, as will be discussed in this chapter, under certain conditions, polyphenols can act as pro-oxidants. This creates an avenue for utilizing polyphenols with a different perspective. In cancer cells, polyphenols may act as pro-oxidants, and this characteristic of these compounds is inspiring researchers to develop strategies so that they can be used either in isolation or in combination with traditional approaches to treat cancer. Knowledge about this phenomenon along with the various new approaches with which this aspect is being utilized by researchers to develop novel strategies for the treatment of cancer will be highly beneficial. Keeping that in mind, this chapter will discuss the role of ROS in the development of cancer and then in subsequent sections, findings from existing literature will be analyzed that will give an idea about the underlying phenomenon and information about a few novel interventions that are being researched upon at current times, utilizing polyphenols for cancer therapeutics.

2. Sources of reactive oxygen species

Sources of ROS generation can be categorized as endogenous or exogenous. When ROS is generated as a by-product of regular cellular metabolism, then it is considered an endogenous pathway of ROS generation. Normally, ROS is generated in electron transport chains, lipid metabolism, purine metabolism, metabolism of amino acids, protein folding, chronic infection or inflammation, or any kind of physical or mental stress. However, if ROS is generated in a cell as a response to an external stimulus (xenobiotics, ionizing radiation, etc.), then the source is termed exogenous [12, 13].

Several enzymes generate ROS as by-products of their activity, and they have a wide range of localization inside cells including the membrane, cytosol, or organelles. The most prominent ROS generating membrane-associated enzyme is nicotinamide adenine dinucleotide phosphate hydrogen (NADPH) oxidase. The organelles involved in ROS production are mainly mitochondria, endoplasmic reticulum, peroxisome, and lysosomes. In mitochondria, apart from the electron transport chain, several enzymes (for example, monoamine oxidase, pyruvate dehydrogenase, α-ketoglutarate dehydrogenase, α-glycerophosphate dehydrogenase, glycerol 3-phosphate dehydrogenase, aconitase, and CYP enzymes) are involved in producing ROS. In the endoplasmic reticulum, xenobiotic metabolizing enzymes of the microsomal monooxygenase family, enzymes involved in the protein folding process (endoplasmic reticulum oxidoreductin-1) and membranous NADPH oxidase are involved in ROS generation. In peroxisomes, the activity of various enzymes (for example, xanthine oxidoreductase, D-aspartate oxidase, acyl CoA oxidases, D-amino acid oxidase, etc.) along with an electron transport chain in the membrane leads to ROS generation. Lysosomes are also a storehouse of ROS. An electron transport chain (involved in the maintenance of high pH concentration) and a high concentration of iron ions assist ROS generation [1, 2, 12].

The major exogenous sources of free radicals are xenobiotics (metals, drugs, cosmetics, food additives, pesticides and insecticides, and polycyclic aromatic hydrocarbons) and radiation. Once cells are exposed to xenobiotics, they are metabolized by various xenobiotic metabolizing enzymes (like cytochrome P-450 and monoamine oxidases) so that they can be readily eliminated from the body. However, in the process, ROS is generated [9].

3. Oxidative stress and effects on major macromolecules

Under normal circumstances, the ROS produced in cells is neutralized or acted upon by a plethora of antioxidant molecules (from endogenous and exogenous sources) which maintains the level of ROS to an optimum level that is required for the proper functioning of cells. However, if the balance between oxidants and antioxidants is perturbed with an overall rise in ROS concentration, oxidative stress is generated. Under conditions of oxidative stress, ROS owing to their reactive nature causes detrimental effects on the major biomolecules of the cell like lipids, nucleic acids, and proteins. The effects of oxidative stress on the major biomolecules of the body are discussed below.

The major effect of ROS on lipids is the oxidation of polyunsaturated fatty acids (PUFA), and the phenomenon is termed lipid peroxidation [14]. Lipid peroxidation is majorly caused by hydroxyl and hydroperoxyl radical and membrane lipids such as glycolipids, phospholipids, and cholesterol are majorly attacked [15]. Generated reactive intermediates undergo further structural modifications to produce toxic aldehydes such as 4-hydoxy-2-nonenal (4-HNE), malondialdehyde (MDA), along with γ-ketoaldehydes [14]. Oxidation of membrane lipids leads to characteristic changes of lipid membrane-like organizational alterations, changes in thermal phase behavior, thickness, symmetry, polarity, and permeability, which might lead to aberration in the interaction between lipids and membrane proteins, ultimately leading to the pathogenesis of diseases [16]. A schematic representation regarding the lipid peroxidation of arachidonic acid, a major component of membrane phospholipids, leading to the formation of MDA (a marker of lipid peroxidation) has been presented in **Figure 1**. Arachidonic acid peroxidation can lead to the formation of several by-products via several pathways. After initial H abstraction by reactive species, the lipid radical generated gives rise to a peroxyl radical in the presence of oxygen. This peroxyl radical in one of its metabolic pathways undergoes subsequent cyclization reactions with simultaneous formation of intermediate radicals, which ultimately give rise to bicyclic endoperoxide. Bicyclic endoperoxides also have several fates, one of which is to disintegrate to form MDA and heptadecatrienoic acid [17].

Reactive oxygen species (ROS) can attack both the sugar and the base of a nucleotide. On reaction with the sugar, the resultant point of abstraction being radicalized leads to cyclization with adjacent base pair leading to the formation of altered nucleosides like 8,5′-cyclopurine-2′-deoxynucleoside [18]. ROS can oxidize DNA bases (both purine and pyrimidines) to form modified bases. Purines are modified to form 2,6-diamino-4-hydroxy-5-formamidopyrimidine, 4,6-diamino-5-formamidopyrimidine, 8-hydroxy adenine, 2-hydroxy adenine, 8-hydroxy guanine, and oxazolone, whereas pyrimidines are oxidatively modified to form 5-hydroxycytosine, cytosine glycol, 5-hydroxydeoxy uridine, uracil glycol, and thymine glycol to name a few [19]. The presence of a modified base pair may lead to mutation, formation of DNA strand breaks, and instability of genetic integrity. A schematic representation of the oxidative modification of guanine to form 8-hydroxy guanine (a marker for oxidative stress in DNA) is represented in **Figure 2A**.

Amino acids such as methionine, proline, cysteine, threonine, tyrosine, histidine, arginine, and lysine are prone to be modified by ROS to produce several compounds including carbonyl derivatives [21, 22]. The formation of carbonylated derivative

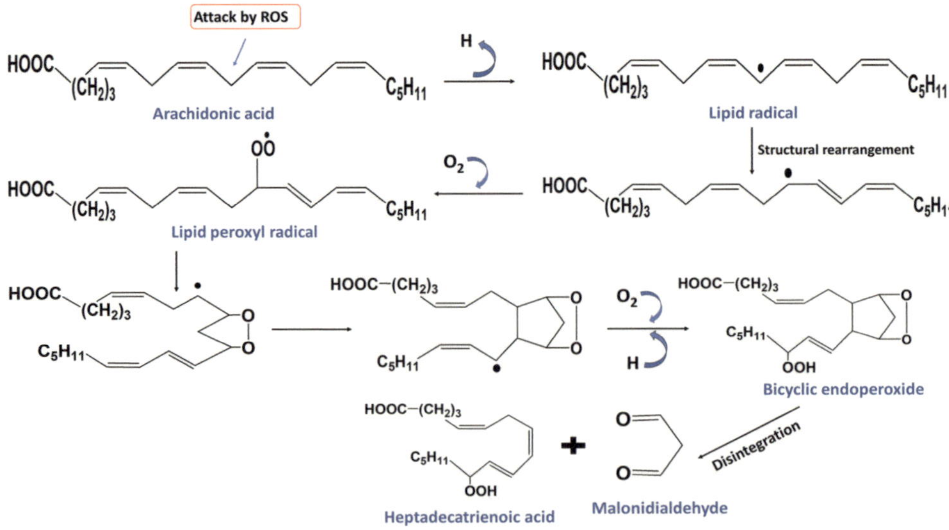

Figure 1.
Lipid peroxidation of arachidonic acid. Adapted from [17].

(a marker for protein oxidation) from arginine has been illustrated in **Figure 2B**. ROS attacks peptides resulting in the formation of carbon-based radicals, which in the presence of oxygen forms peroxyl radical that can undergo further reactions to give rise to different types of by-products. In one pathway, it abstracts hydrogen to form hydroperoxide and subsequently alkoxyl radical on reduction. Alkoxyl radicals can disintegrate to form several by-products. However, in the absence of oxygen, the initial alkyl radical forms cross-link with another adjacent carbon-based radical to form aggregates [23]. A schematic illustration of oxidative stress-mediated damage to peptide chains has been presented in **Figure 3**. Attack by ROS brings about effects

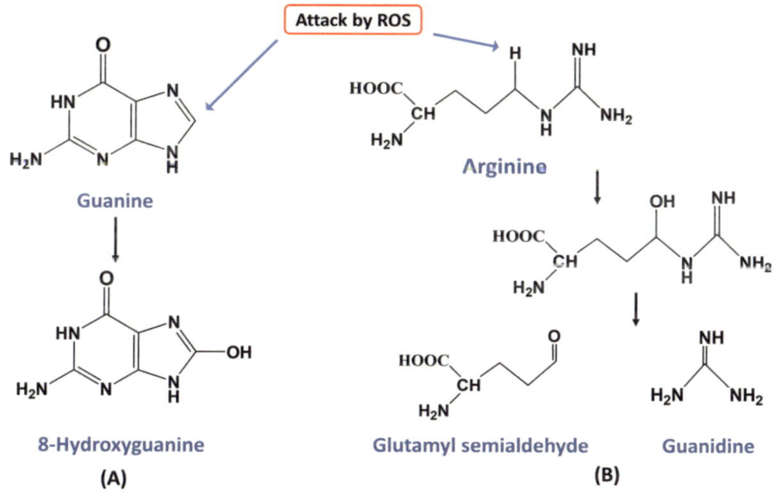

Figure 2.
Schematic representation for (A) Reactive oxygen species-mediated modification of guanine to 8-Hydroxyguanine (adapted from [18]); (B) Reactive oxygen species-mediated oxidation of arginine to form carbonylated by-products (adapted from [20]).

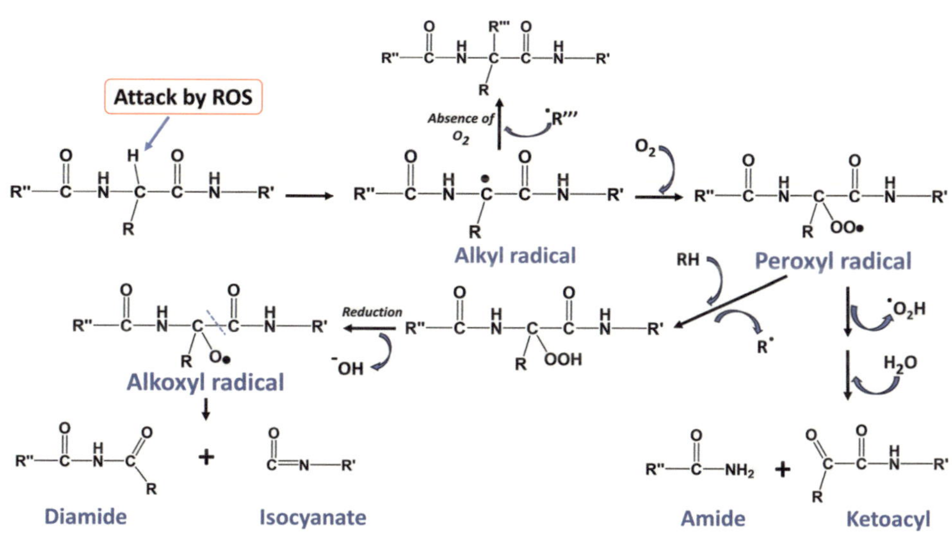

Figure 3.
Oxidative damage of peptide backbone by reactive oxygen species. Adapted from [23, 24].

such as structural modification, denaturation, aggregation of proteins, and loss/abnormal functioning of the protein and enzymes.

4. Role of oxidative stress in the pathogenesis of cancer

Carcinogenesis is a multi-step process in which genetic alterations, oncogene activation, tumor suppressor gene deactivation, epigenetic modifications, and subsequent perturbations of signal transduction pathways bring about uncontrolled and abnormal growth, proliferation, and differentiation of cells [25]. Oxidative stress is found to be involved in the genesis of several types of cancer [26]. ROS-induced modifications in nucleic acids may cause mutation of genes and damage to DNA strands. These in turn can lead to catastrophic events like oncogene activation and tumor suppressor gene deactivation, which are primary events leading to carcinogenesis.

Oxidative stress can also modulate the signal transduction pathways of a cell by either directly modifying a key molecule that initiates a pathway or by activating/deactivating a molecule which in turn activates a signal transduction pathway. Signaling pathways activated by ROS such as mitogen-activated protein kinase (MAPK) pathway, phosphatidylinositol-4,5-bisphosphate 3-kinase-protein kinase B (PI3K-AKT) pathway, Kelch-like ECH associating protein 1-nuclear factor erythroid 2-related factor 2 (Keap 1-Nrf2) pathway, Janus kinase-signal transducers and activators of transcription (JAK-STAT) pathway, wingless-related integration site-β-catenin (Wnt-β-cat) pathway, p53 pathway, SRC pathway (non-receptor tyrosine kinase initiated pathway), and nuclear factor kappa-light-chain-enhancer of activated B cells (NF-κB) pathway lead to the activation of several downstream targets like transcription factors which promote cellular proliferation and survival, inflammation, epithelial-mesenchymal transition (EMT), invasion and angiogenic capacity [27, 28]. Structural alteration via oxidation of crucial amino acid residues in receptors (leading to activation of a pathway even in the absence of ligand), downstream kinases and inhibitor proteins (that otherwise keep the key signal transducer in an inactive form) may be involved behind erroneous activation of signaling pathways [29].

Keap-1-Nrf2 pathway is activated when ROS oxidatively deactivates Keap-1, which under normal conditions keeps the transcription factor Nrf-2 in an inactive form. Once activated, Nrf2 moves to the nucleus and activates transcription of oxidative stress-responsive genes, which upregulate the endogenous antioxidant level of the cell [30, 31]. In cancer cells, ROS levels are higher than in normal cells due to chronic inflammation and altered metabolism. In the presence of excessive ROS beyond a certain level, programmed cell death mechanisms are initiated. To maintain survival, cancer cells constitutively express an antioxidant defense system [30]. The PI3K-AKT pathway is normally inhibited by the phosphatase and tensin homolog (PTEN) by dephosphorylation of PI3K. ROS has been found to inhibit PTEN activity by forming a disulfide bridge between cysteine residues, leading to the activation of the pathway [32]. AKT regulates the expression of an array of cellular proteins supporting cell survival and proliferation, metabolism, and angiogenesis [33]. MAPKs are a family of kinases with members like extracellular signal-regulated kinase (ERK), c-Jun N-terminal kinase (JNK), and p38, which are involved in the mediation of life processes such as cellular proliferation, differentiation, apoptosis, and immunological activities [27]. They are majorly activated when a signaling ligand binds to its receptor, but ROS can activate the receptor even in the absence of the ligand which leads to the activation of the signaling cascade [34]. Initiation of signal transduction pathways by ROS leads to the activation of several transcription factors like NF-κB, hypoxia-inducible factor-1α (HIF-1α), activator protein- 1(AP-1), p53, STAT-3, Nrf-2, etc. [27, 28, 35]. Apart from activation by signaling pathways as downstream effectors, ROS can activate both NF-κB and HIF-1α by deactivating their inhibitors directly. ROS can also modify the inhibitory component of NF-κB, inhibitors of NF-κB (IκBs), or activate NF-κB-inducing kinase (NIK), thereby leading to activation of the transcription factor [27, 36]. Activated NF-κB is involved in the induction of genes for cellular survival and proliferation, inflammatory mediators like cytokines and chemokines, invasion and metastatic potential, angiogenesis, and cellular metabolism and also mediates the induction of changes at genetic and epigenetic level in cells, thereby setting the base for carcinogenesis [37]. HIF-1α is activated by ROS via oxidation-based deactivation of its inhibitor, prolyl hydroxylase domain protein 2 (PHD2) [38]. Activated HIF-1α is involved in the transcription of genes that support angiogenesis, metabolism, cellular survival, and proliferation, which augments carcinogenesis [39]. AP-1 on being activated regulates expression of genes related to cellular growth and death, angiogenesis, and upregulated metastatic potential [40]. Transforming growth factor-β (TGF-β) is a crucial cytokine that promotes EMT (via transcription factors like Snail and Slug), angiogenesis, and evasion of immune response in cancer cells. In normal cells, TGF-β exhibits a tumor-suppressor role, but in transformed cancer cells, it manifests pro-carcinogenic activity. It is normally secreted by cells in a latent form, associated with a latency-associated peptide (LAP), and is sequestered in the ECM. On being activated by any form of stimulus, it binds to its receptor and initiates either canonical (SMAD-mediated) or non-canonical pathways (MAPK, PI3-AKT, etc.). A dynamic interplay occurs between ROS and TGF-β in cancer cells where ROS not only increases the expression of TGF-β but also diffuses out of the cell and activates latent TGF-β by modifying the associated LAP domain. In turn, TGF-β acts to increase the ROS content in cells via both direct ROS generation (mediated by NOX-4) or by downregulating the cellular antioxidant status. Moreover, ROS also mediates several pathways induced by TGF-β leading to EMT, like the NF-κB pathway [41, 42]. Therefore, several factors and phenomena induced by ROS act in synergy to induce carcinogenesis in cells. A schematic representation of oxidative stress generation and its carcinogenic effects in a cell has been presented in **Figure 4**.

Another important aspect regarding ROS is its ability to modulate the tumor microenvironment (TME) that promotes tumor aggressiveness and cancer metastasis. Tumor microenvironment is composed of several types of cells—cancer-associated fibroblasts (CAFs), tumor-infiltrating leukocytes (like myeloid-derived suppressor cells (MDSCs), tumor-associated macrophages (TAMs) and T cells), mesenchymal stem cells, endothelial cells, lymph and blood vessels, and extracellular matrix (ECM) components [43]. ROS, inflammatory cytokines, and chemokines released by cancer cells trigger infiltration of immune cells to the tumor site and modulate characteristics of cells in the immediate environment. Infiltrated immune cells further aggravate the oxidative stress load of TME via the activity of their intracellular pro-oxidant enzymes like NOX, MPO, etc. [44]. ROS (like H_2O_2) released by cancer cells can diffuse into the extracellular environment and modulate the differentiation of CAFs into subtypes like myofibroblasts, which can modulate TME and promote cancer. Myofibroblasts are a type of CAF that secrete ECM modulators and can therefore promote cellular migration [45]. Cancer cell-generated ROS drives stress generation in CAFs and metabolically re-wires them to switch to a glycolytic mode of energy generation and undergo autophagy and subsequently provide substrates like pyruvate and lactate to cancer cells for enhanced survival [46].

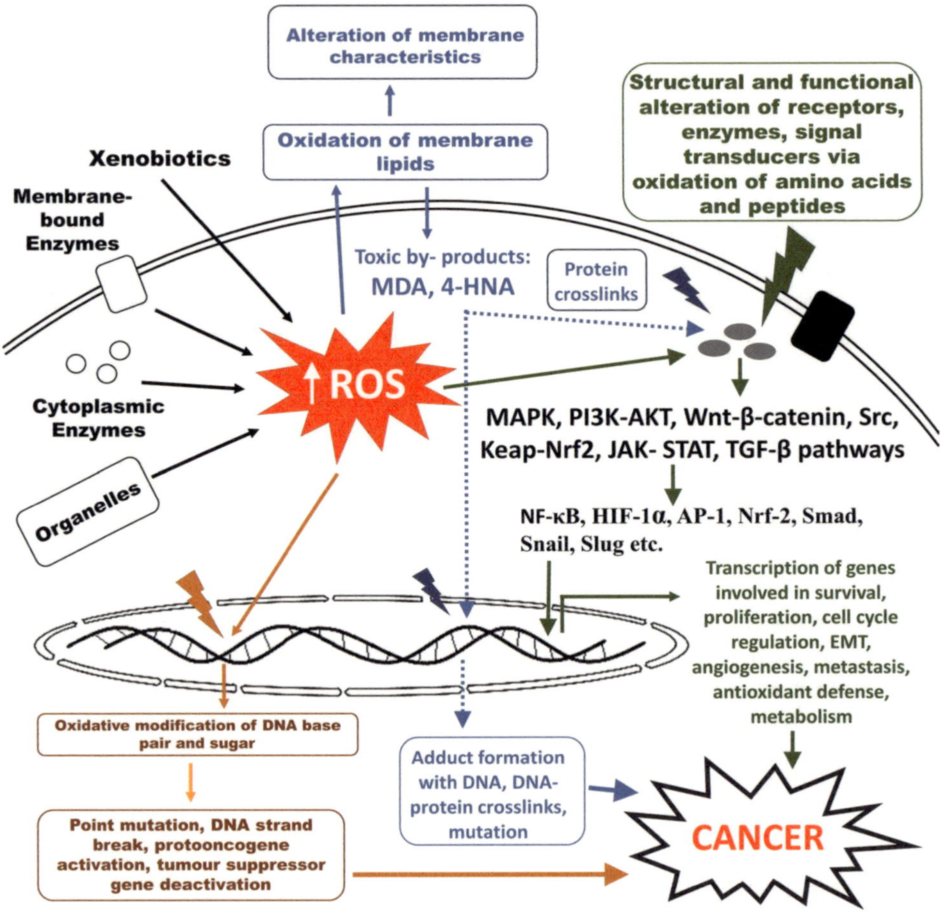

Figure 4.
Schematic illustration depicting the carcinogenic effects of reactive oxygen species and its sources in a cell.

Immune cells summoned by tumor cells into TME act to suppress immunoreactivity against tumor, which facilitates cancer metastasis and promotes resistance to therapy [47]. ROS has been found to impede cytokine production by T cells and induce their hypo-responsivity [48, 49]. ROS in TME regulates the immunosuppressive activity by T_{reg} cells. Under conditions of microenvironmental oxidative stress, T_{regs} undergo apoptosis and the apoptotic T_{regs} secrete large amounts of ATP and convert it into adenosine with the help of membrane molecules. Adenosine binds to A_{2a} receptors in nearby T cells and suppresses their cytokine production leading to subdued T-cell reactivity [50]. ROS generated by tumor cells via the activity of NOX-4 facilitates TAM polarization to pro-tumorigenic phenotype (M2), which facilitates malignancy [51]. In response to oxidative stress generated by tumor cells, activated immune cells too generate oxidative stress intracellularly, which not only induces their differentiation into pro-carcinogenic phenotype but also influences the activity of other immune cells in the vicinity. ROS generated by MDSCs suppresses the proliferation and function of T cells and promotes cancer cell growth [52, 53]. In TME, immature myeloid cells are induced to produce ROS that in turn restricts their differentiation into mature immune cells like macrophages, dendritic cells, and granulocytes [54]. High levels of ROS generation in TAMs upregulate the secretion of tumor necrosis factor-α that facilitates cancer metastasis [55]. When cultured in tumor cell-conditioned media, activation of myofibroblasts was triggered via enhanced intracellular ROS generation [56]. Therefore, a dynamic interplay exists between cancer cells and the surrounding cells in TME (both immunological or non-immunological), where ROS plays a significant role as a mediator. ROS acts both as an initiator and perpetuator of cancer and, as we will discuss in subsequent sections, can also be employed in cancer therapeutics. A minimal amount of ROS is required for normal cell functioning, whereas a high level of oxidative stress leads to carcinogenesis as already discussed. However, further increase in ROS level beyond a threshold triggers cell death pathways like apoptosis and autophagy [27].

5. Polyphenols: potential candidates as chemotherapeutics

Polyphenols are plant-based active compounds that are originally secondary metabolites of plants, present in their fruits, leaves, and seeds. As the name suggests, their structure characteristically contains phenol rings with specific substitutions at various positions. Polyphenols are a huge family of phytochemicals with more than 8000 known compounds [57]. From the functional point of view, polyphenols have been found to have antioxidant and anti-inflammatory effects on cells and thereby can be used as a preventive agent against the development of disorders arising from the state of chronic inflammation like cardiovascular disease, neurodegenerative disorders, atherosclerosis, diabetes, hypertension, and cancer to name a few [58, 59].

On the basis of their structure, polyphenols are broadly categorized into four major groups, namely, flavonoids, phenolic acids, stilbenes, and lignans [60]. Apart from these, a category of polyphenols encompasses miscellaneous members like curcuminoids (http://www.phenol-explorer.eu).

5.1 Polyphenols as pro-oxidants

Just like the two sides of a coin, polyphenols in addition to their antioxidant capacity also act as pro-oxidants under conditions of high concentration and pH and in the presence of transition metal ions (Cu^{2+}/Fe^{3+}) [61, 62]. Polyphenols reduce metal ions and themselves get converted to a phenoxyl radical. In the presence of molecular oxygen, the phenoxyl radical gets converted to a quinone, and a superoxide anion is generated. This superoxide anion then forms hydrogen peroxide. Hydrogen peroxide in the presence of transition metal ions forms hydroxyl radicals via Fenton reactions or Haber-Weiss reaction in the presence of superoxide anion [61]. Several *in vitro* studies have found that polyphenolic compounds caused DNA damage in cells when transition metals are present [63–66]. Polyphenols reduce copper ions (Cu^{2+}) bound to chromatin, and it leads to the generation of ROS, as discussed above, in very close vicinity to DNA molecules, and the ROS generated thereby causes DNA damage [64]. Cancer cells, owing to their metabolically re-programmed condition and chronic inflammatory status, survive in a state containing higher oxidative stress, altered pH level, and high concentration of transition metal ions [67, 68]. As a high ROS level beyond a certain threshold level activates signaling pathways leading to programmed cell death, the pro-oxidative feature of polyphenols in cancer cells further increases the oxidative stress load in them, leading to the induction of cell death. This characteristic is exploited in the usage of polyphenols as chemotherapeutic agents. A schematic diagram of pro-oxidant activity of phenols in high oxidative stress and metal ion concentration in cancer cells has been illustrated in **Figure 5**.

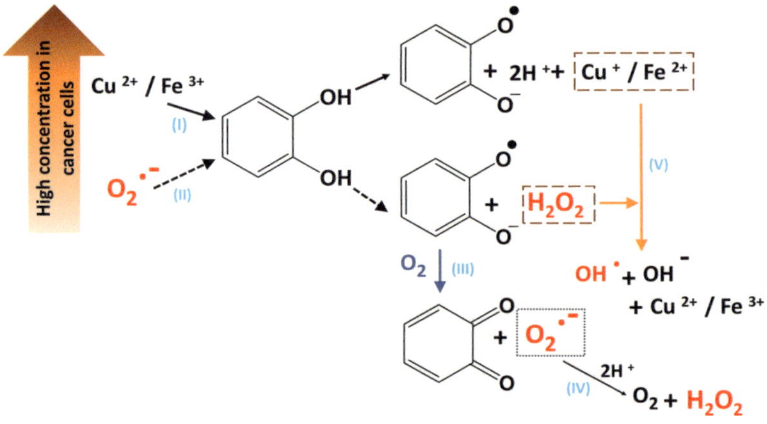

Figure 5.
Pro-oxidant activity of phenols in cancer cells under conditions of high oxidative stress and transition metal ion concentration. (I) Phenols react with metal ions to form phenoxyl radical and reduced metal ions, (II) Phenol reacts with superoxide anion to form phenoxyl radical and hydrogen peroxide, (III) Phenoxyl radical in presence of molecular oxygen undergoes further oxidation and forms quinone and superoxide anion, (IV) Superoxide anion undergoes spontaneous dismutation in presence of cellular proton to form hydrogen peroxide and oxygen, (V) Reduced metal ion and hydrogen peroxide react to generate hydroxyl radical. Adapted from [61].

Several phenomena lead to the generation of ROS by flavonoids and it includes their transition metal or pH-dependent auto-oxidation, oxidation in the presence of both molecular oxygen and transition metals, oxidation of cellular NADH (nicotinamide adenine dinucleotide hydrogen) and antioxidants by the intermediate phenoxyl radicals, and impairment of mitochondrial respiration [69]. Pro-oxidant activity is also a function of the concentration and structure of flavonoids. At higher concentrations, flavonoids exert a pro-oxidant effect instead of acting as an antioxidant [70]. The number and position of hydroxyl groups in the second phenol ring and the occurrence of the double bond in the pyran ring are important for pro-oxidant activity of flavonoids [61].

The direct pro-oxidant effect of phenolic acids is prominent in the presence of transition metal ions and is also a function of their structure and oxidation potential. In the presence of Cu^{2+} ions, hydroxycinnamic acid derivatives (caffeic acid, chlorogenic acid, sinapic acid, and ferulic acid) induced oxidative stress-mediated DNA damage. The presence of orthohydihydroxyl group and 4-hydroxyl-3-methoxl group in the compounds markedly enhanced their pro-oxidant functionality. Monohydroxylated derivatives like 3-hydroxycinnamic acid and 4-hydroxycinnamic acid did not depict DNA damaging activity even in the presence of Cu^{2+} ions. Moreover, compounds with lower oxidation potential were found to have greater pro-oxidant ability [71]. Structural dependence of phenolic acids for pro-oxidant ability was also observed in the study conducted by Khan and Hadi, where it was found that the DNA cleaving potential of syringic acid was lower than that of gallic acid due to methylation of the hydroxyl groups, whereas decarboxylation of gallic acid leads to enhancement of DNA damaging potential as is observed in pyragallol [72]. Rosmarinic acid has also been found to induce oxidative stress-mediated DNA damage in the presence of transition metals (Cu/Fe) [73]. Tannic acid, ellagic acid, and gallic acid were found to have antioxidant activity in the Chinese hamster fibroblast cell line, but the effect was reduced in the presence of hydrogen peroxide and Cu^{2+} ions. Moreover, these phenolic acids induced thiol oxidation, protein carbonylation,

DNA damage, and apoptosis in tested cell lines individually (at higher concentrations) and the effect was more conspicuous in the presence of Cu^{2+} ions [74, 75].

In the case of stilbenes too, structure, concentration, and presence of transition metals play an important part in facilitating pro-oxidant effects. Resveratrol, oxyresveratrol, and piceatannol (hydroxylated derivatives of trans-stilbene) caused ROS generation and ds break in plasmid DNA in the presence of Cu^{2+}, whereas this effect was not observed in the case of trans-stilbene. Moreover, the DNA-damaging activity was observed only at higher concentrations of oxyresveratrol [65, 76]. Resveratrol and piceatannol have also been observed to induce DNA damage in lymphocytes in the absence of added Cu^{2+}, possibly via mobilization of intracellular Cu^{2+}. However, in that case, much higher dosage of the compounds was required to be administered [77]. Structure-activity analysis study conducted with resveratrol and seven of its synthetic structural analogs has found that the presence of orthodihydroxyl group or 4-hydroxyl group containing compounds has higher pro-oxidant activity along with the enhanced capacity for inducing DNA damage [78].

6. Polyphenols in cancer therapeutic research: a brief overview of select few

Polyphenols exhibit antioxidant and anti-inflammatory effects, which accentuate their chemopreventive roles. However, they also act as pro-oxidants in cancer cells because of the altered cellular environment, an attribute that is harnessed for facilitating anti-tumorigenicity or anti-carcinogenicity in chemotherapeutics. In recent times, novel chemotherapeutic strategies are being designed to incorporate polyphenols into mainstream treatment regimens either as an individual component or as an adjunct with traditional chemotherapeutic drugs like doxorubicin, paclitaxel, etc. In this section, a few polyphenols will be discussed which are known to have potent chemotherapeutic properties, whose structures (along with the basic structures of the major groups of polyphenols) have been depicted in **Figure 6**.

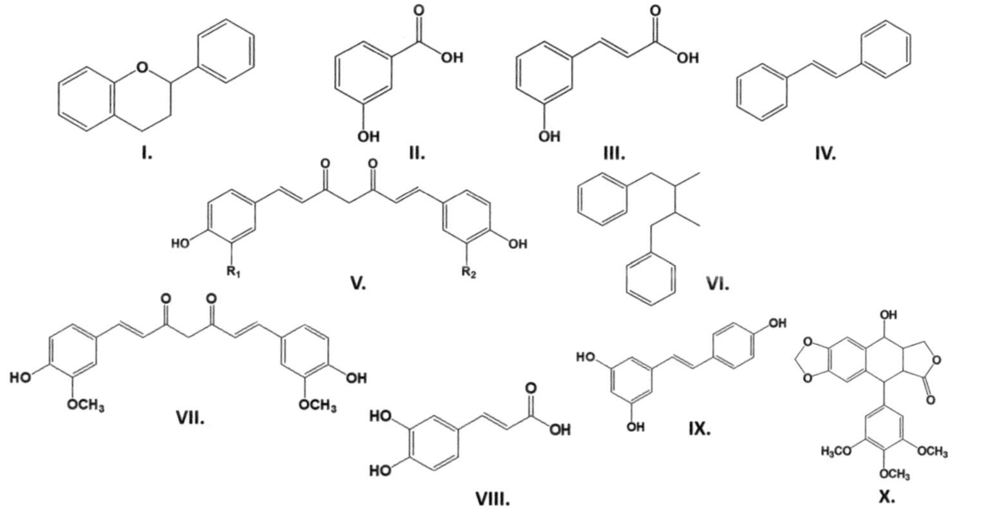

Figure 6.
(I–VI) Basic structures of polyphenols. I. Flavonoids; II–III., Phenolic acids, Hydroxybenzoic acid, and Hydroxycinnamic acid, respectively; IV. Stilbene; V. Curcuminoid (R1, R2: H/OCH3); VI. Lignan. (VII– X) Structures of discussed polyphenols. VII. Curcumin; VIII. Caffeic acid; IX. Resveratrol; X. Podophyllotoxin.

6.1 Curcumin

Curcumin (diferuloylmethane, Cur), a curcuminoid, is obtained from the rhizome of *Curcuma longa*, generally called turmeric. Cur can act as an antioxidant molecule by virtue of its structure (the β-diketo group, carbon-containing double bonds, and phenyl ring with hydroxyl and methoxy substituents). The antioxidant role of Cur in normal cells protects it from ailments resulting from oxidative stress. In a study conducted by Dange et al., it was observed that feeding mice Cur (post-irradiation) prevented the development of radiation-induced thymic lymphoma [79]. However, in radio-resistant pancreatic cancer cell line, Cur was found to induce radio-sensitivity leading to DNA damage induction, cell cycle arrest, and apoptosis on irradiation [80]. This contrasting role of Cur provides evidence supporting its differential activity in normal and cancerous cells. Moreover, the dosage of Cur used in therapeutic procedures may also decide the pro- or anti-oxidant role of Cur [81].

Several *in vitro* studies have observed a pro-oxidant role of Cur in the presence of Cu^{2+} ions, and the generated ROS led to DNA damage and cellular death [63, 82]. The presence of a conjugated beta-diketone moiety in the structure of Cur is speculated to be accountable for the pro-oxidant effect [63]. As presented in **Table 2**, Cur has shown toxicity in cancer cells originating from various organs by increasing the oxidative stress level. Cur induces senescence or cell death in exposed cells in the form of apoptosis, autophagy, ferroptosis, and pyroptosis by increasing the level of oxidative stress in them. Both the mitochondrial dysfunction-mediated intrinsic pathway and Toll-like receptor-mediated extrinsic pathway of apoptosis have been found to be induced by Cur, based on the type of cancer cell [87]. Some mechanisms by which it mediates pro-oxidative anticancer effects include the downregulation of cellular antioxidant enzymes, induction of damage to DNA and cellular lipids, induction of

Compound	Test model	Effect	Ref.
Curcumin	Osteosarcoma cell lines	ROS generation, anti-proliferative, cell cycle arrest, apoptosis/ ferroptosis, anti-invasive	[83]
	Mouse model xenografted with cancer cells	Anti-tumorigenic	
	Non-small-cell lung cancer cell line	ROS generation, anti-proliferative, DNA damage, G2/M cell cycle arrest, apoptosis, inhibits colony formation	[84]
	Hepatocellular carcinoma cell line	ROS generation, anti-proliferative, apoptosis/pyroptosis	[85]
	Colorectal cancer line	ROS generation, anti-proliferative, anti-migratory, apoptosis/ senescence, activation of ROS-Nrf2-miR 34a/b/c signaling cascade. Potentiated activity of 5-fluorouracil	[86]
	Mouse xenograft model	Anti-tumorigenic	
	Liver cancer cell line	ROS generation, anti-proliferative, apoptosis	[87]
	Lung cancer cell line (chemo-resistant)	ROS generation, anti-proliferative, apoptosis, p38 activation	[88]
	Cervical cancer cell line	ROS generation, anti-proliferative, G2/M cell cycle arrest, apoptosis, autophagy, cellular senescence (via p53-p21 induction)	[89]

Compound	Test model	Effect	Ref.
CUR and Cisplatin: Nano-Liposome encapsulated	Hepatocellular carcinoma cell line	ROS generation, anti-proliferative (synergistic effect), apoptosis (synergistic effect), ERK, and p53 activation	[90]
	Mice model with xenografted tumor	Increased circulation and residence time, anti-tumorigenic, increased bioavailability, reduced toxicity.	
WZ26 (CUR analog)	Cholangiocarcinoma cell line	More potent activity than Cur itself, ROS generation, anti-proliferative, inhibits migration, apoptotic, G2/M cell cycle arrest, inhibits STAT3 signaling	[91]
	Mice model with xenografted cancer cells	Anti-tumorigenic	
CUR in combination with Cisplatin	Papillary thyroid cancer cell lines	ROS generation, anti-proliferative, apoptotic, anti-migratory, inhibition of matrix metalloproteinases, synergistic effects with cisplatin, decrease stemness of cancer stem cells, inhibit JAK-STAT3 pathway	[92]
	Breast cancer cell line	ROS generation, inhibition of DNA repair protein, sensitized resistant triple negative breast cancer cells to carboplatin, apoptosis	[93]
Caffeic acid (CA)	Colon cancer cell line	ROS generation, anti-proliferative, apoptosis	[94]
	Breast cancer cell lines	ROS generation, apoptosis	[95]
	Chronic myeloid leukemia cell line	ROS generation, anti-proliferative, apoptosis, activation of transglutaminase type 2	[96]
CA N-butyl ester	Lung cancer cell line	ROS generation, DNA damage, anti-proliferative, necrosis	[97]
CA-phenyl ester	Myeloma cell line	Oxidative stress, DNA damage, anti-proliferative, apoptosis	[98]
	Normal peripheral blood cells	Non-toxic	
	Medulloblastoma cell line	ROS generation, anti-proliferative, apoptosis, anti-invasive, inhibition of NF-κB Radio-sentitization	[99]
CA-3,4-dihydroxy-phenethyl ester	Breast cancer cell lines	ROS generation, anti-proliferative, apoptosis	[100]
CA-Bortezomib-Iron-conjugated nanomedicine	Colon carcinoma cells (Mice)	ROS generation, DNA damage, Inhibit NF-κB. Complement apoptotic action of the drug	[101]
	Normal fibroblast cell	Nominal toxicity	
CA, TRF, nano-emulsion Cisplatin	Lung cancer cell line Hepatocellular carcinoma cell line	ROS generation, apoptosis. Complements the chemotherapeutic action of Cisplatin by reducing the dosage required and toxicity	[102]
	Normal human embryonic kidney cells	Lowered toxicity. Viability significantly high in comparison to only cisplatin treatment	
Poly CA-Bortezomib-Folic acid coated Au-NP nanomedicine	Squamous cell carcinoma cell line	ROS generation, anti-proliferative, apoptosis, inhibits NF-κB	[103]
	BALB/c nude mice xenografted with cancer cells	Anti-tumorigenic	

Compound	Test model	Effect	Ref.
Resveratrol (RSV)	Cervix cancer cell lines	ROS generation, anti-proliferative, autophagy/mitophagy	[104]
	Normal cell line	Non-toxic (very high IC_{50})	
	Melanoma cell line	ROS generation, ER stress, cell cycle arrest, anti-proliferative, apoptosis, Induction of p38 and p53	[105]
	Colon cancer cell line	ROS generation, anti-proliferative, autophagy, apoptosis	[106]
	Ovarian cancer cell lines	ROS generation, apoptosis, inhibit Notch1 and AKT	[107]
	Anaplastic thyroid cancer cell line	ROS generation, apoptosis	[108]
	Prostate cancer cells (Mice)	ROS generation, anti-proliferative, anti-invasion, apoptosis, induction of p53 and HIF-1α	[109]
	Pancreatic cancer cell lines	ROS generation, apoptosis (downregulated NAF-1), induction of Nrf-2 Sensitize cells non-responsive to Gemcitabine	[110]
	Breast cancer cell lines Lung cancer cell line	ROS generation, DNA damage, anti-proliferative, senescence (via induction of deleted in liver cancer-1)	[111]
	Normal breast epithelial cell line	Non-toxic	
	Cancer stem cells derived from the human ovarian cancer cell line	ROS generation, apoptosis. Reduced Sox, Nanog expression in surviving cells. Reduced cell renewal capacity in surviving cells	[112]
	Human normal fibroblast	Non-toxic	
RSV-nanomedicine *,#	Colorectal cancer cells	ROS generation, anti-proliferative, lipid peroxidation, reduced invasive capability, ferroptosis	[113]
	Mice (nude) xenografted with colorectal cancer cells	Anti-tumorigenic	
Podophyllotoxin (PT)	Colorectal Cancer Cells	ROS generation, ER stress, G2/M cell cycle arrest, anti-proliferative, inhibition of anchorage-independent growth, p38 activation, apoptosis	[114]
PT acetate	Non-small cell lung cancer	Radio-sensitization, increased apoptosis in combination with radiation, ROS generation, inhibits ERK, activates p38	[115]
	Mice (nude) xenograft model	Anti-tumorigenic	
$Ptox^{DPT}$	Hepatocellular carcinoma cell lines	ROS generation, anti-proliferative, anti-invasion, inhibit matrix metalloproteinases, expression, downregulate expression of vimentin, Snail and Slug, upregulate expression of E-cadherin and p53, inhibition of PI3K/AKT/mTOR pathway	[116]
PT-based PPM-DDS with cucurbitacin B (micelles)	Multi-drug-resistant lung cancer cell lines Mice (nude) model xenografted with cancer cells	ROS generation, anti-proliferative, anti-tumorigenic	[117]

Compound	Test model	Effect	Ref.
DPT	Prostate cancer cell lines (chemoresistant)	ROS generation, G2/M cell cycle arrest, apoptosis	[118]
	Non-small cell lung cancer cells (Gefitinib-resistant)	ROS generation, ER stress, G2/M cell cycle arrest, anti-proliferative, inhibit anchorage-independent growth, apoptosis. Chemo-sensitization by reducing activity of receptor tyrosine kinases (EGFR and MET and downstream kinases ERK and AKT)	[119]
DPMA	Non-small cell lung cancer cells	ROS generation, hamper microtubule formation, induce p53 expression, apoptosis	[120]
	Mesenchymal stem cells	Lower level of toxicity	
	Human umbilical vein endothelial cells	Anti-angiogenic	
PPT	Esophageal Squamous Cell Carcinoma Cell line	ROS generation, ER stress, G2/ M cell cycle arrest, anti-proliferative, inhibit anchorage independent growth, apoptosis, activation of JNK-p38 pathway	[121]
	Colorectal Cancer Cells	ROS generation, ER stress, cell cycle arrest, anti-proliferative, apoptosis, p38 activation	[122]

*Abbreviations (not mentioned in text): **WZ26**: (1E,4E)-1-(3-bromo-4-hydroxyphenyl)-5-(4-hydroxy-3-methoxyphenyl)penta-1,4-dien-3-one; **TRF**: tocotrienol; **ER**: endoplasmic reticulum; **NAF-1**: nutrient-deprivation autophagy factor-1; **PPM-DDS**: polymeric prodrug micellar-based drug delivery systems; *: RSV-loaded poly (ε-caprolactone)-poly (ethylene glycol) (PCL-PEG) nanoparticles (NP) encapsulated with red blood cell (RBC) membrane (RSV-NPs@RBCm); #: RSV-NPs@RBCm co-delivered with a tumor-penetrating peptide iRGD (RSV-NPs@RBCm&iRGD); **PtoxDPT**: combinatorial compound of 4′-demethylepipodophyllotoxin and dithiocarbamate (di-2-pyridineketone hydrazone dithiocarbamate S-propionate podophyllotoxin ester); **DPT**: deoxypodophyllotoxin (analog of PT); **DPMA**: 2,6-dimethoxy-4-(6-oxo-(5R,5aR,6,8,8aR,9-hexahydrofuro[3′,4′:6,7]naphtho[2,3-d][1,3]dioxol-5-yl)phenyl ((R)-1-amino-4-(methylthio)-1-oxobutan-2-yl) carbamate (derivative of DPT); **EGFR**: epithelial growth factor receptor; **MET**: mesenchymal epithelial transition; **PPT**: picropodophyllotoxin (epimer of PT).*

Table 2.
Compilation of studies showing chemotherapeutic effects of polyphenols and their derivatives alone or in combination with traditional anticancer drugs.

mitochondrial dysfunction, inhibition of DNA repair proteins, induction of pro-apoptotic proteins, and repression of anti-apoptotic proteins, modulation of the activity of cell cycle regulatory molecules and pathways like JAK-STAT3, ERK, p38, and p53. Cur has also been found to be suitable as an adjunct to traditional chemotherapeutic drugs. It acts synergistically with the drugs to induce anti-tumorigenicity and sensitizes chemo-resistant cells to chemotherapeutic drugs [92, 93].

Due to constraints like its poor solubility in water and low bioavailability, its concentration level in plasma does not reach the values that are required for it to be a potential therapeutic agent. Hence, newer drug delivery systems are being designed like metal-organic frameworks, stimuli-sensitive nanocarriers with added ligands for target-specific homing of drugs, or encapsulation in nano-sized liposomes, which carry curcumin alone or in combination with other chemotherapeutic drugs. These techniques lead to enhanced drug retention, improved drug targeting to tumor site, better uptake by tumor cells, and higher efficacy in inducing anti-tumorigenic response along with reduced toxicity [90, 123, 124]. Moreover, several analogs of Cur are either being isolated or synthesized, e.g., terpene-conjugated analogs (bisabolocurcumin ether, demethoxybisabolocurcumin ether) and WZ26, which have shown better potential as anti-cancer agents in comparison to Cur [91, 125].

6.2 Caffeic acid

Caffeic acid (3, 4-dihydroxycinnamic acid, CA) is a hydroxycinnamic acid derivative and is found in a variety of vegetables, fruits, and beverages like coffee. Several compounds are derived from CA including its phenyl-, methyl-, ethyl-, butyl-, and benzyl ester, chlorogenic acid, methyl caffeate, methyl dihydrocaffeate, octyl caffeate, ferulic acid, and cichoric acid. CA is a potent antioxidant involved in the direct scavenging of reactive species [126]. However, a high concentration of caffeic acid in the presence of transition metal ions can lead to the generation of its pro-oxidant nature [127]. As already mentioned in the previous section, CA can act as a pro-oxidant when transition metal ion like Cu^{2+} is present and the ROS generated thereby induces DNA damage [71]. A low level of ROS generation by CA might also facilitate their pro-survival or antioxidant effects in normal cells by induction of pro-survival pathways like ERK [128].

The potential of both CA and its derivatives has been investigated in several studies and it has been found that in cancerous cells, it exerts a pro-carcinogenic effect, whereas in normal cells, they are either non-toxic or have nominal toxicity (**Table 2**). Therefore, they have a selective action profile which asserts their candidature as potent anti-cancer agents. The mechanisms by which CA and its derivatives induce pro-oxidative anticancer effects mainly include induction of oxidative stress by either ROS generation and/or depletion of cellular antioxidants, DNA damage, apoptosis, mitochondrial dysfunction, and inhibition of pro-survival pathways like NF-κB (data compiled in **Table 2**).

Caffeic acid (CA) as nanomedicine or as nano-emulsions has been tested with chemotherapeutic drugs like bortezomid and cisplatin [102, 103]. CA-Bortezomid not only have selective toxic effects on cancer cells, these have also shown anti-tumorigenic effects in mice cancer model that has been xenografted with tumor cells [103]. Chemotherapeutic drugs have severe side effects and sometimes cancer cells become unresponsive to treatment. Such conjugated nano-medicines overcome this problem, resulting in increased uptake of drugs, lessening the amount of effective dosage, and diminishing systemic toxic responses. CA was observed to be effective in inducing cell cycle arrest and apoptosis in chronic myeloid leukemia cells, which are resistant to Imatinib mesylate [129]. In addition to CA, its natural derivatives have also been found to induce anticancer effects in cancer cell lines. Therefore, there is scope in areas of active drug designing too, where novel CA derivatives can be designed and newer anticancer treatment options can be investigated.

6.3 Resveratrol

Resveratrol (trans-3,4′,5-trihydroxystilbene, RSV) is the most abundantly studied component of the stilbene family of polyphenols. RSV has been extensively studied as an anticancer molecule in cancer cell lines and in *in vivo* models (**Table 2**). It selectively induces toxicity in cancer cell lines and is non-toxic in normal cells. The mechanisms by which RSV induces pro-oxidative anticancer effects include induction of oxidative stress either by depleting the existing reserve of antioxidant enzymes or by inhibiting their activity, induction of mitochondrial dysfunction and endoplasmic reticulum stress, induction of cell cycle arrest by inducing changes in expression level of cell cycle regulatory molecules, causing interference in pathways involved in generation of energy (glycolysis and oxidative phosphorylation) [104], modulation of signaling pathways involving transcription factors and kinases like p38, p53, NF-κB, HIF-1α, Sox, Nanog, AKT, and Notch signaling, induction of apoptosis, autophagy, ferroptosis, and cellular senescence (data compiled in **Table 2**).

Studies have shown that RSV reduces the self-renewal ability of cancer stem cells and even sensitizes resistant cells to chemotherapeutic drugs like gemcitabine [110, 112]. Development of nano-medicines using RSV is also in the process, where RSV-containing nanoparticles are enabled to escape cellular immune defense and more efficiently reach the target site [113]. These nanoparticles have also been found to efficiently execute their anti-tumorigenic activity in mice model. In a few studies conducted in pancreatic cell line, it was observed that RSV counteracted hypoxia and hyperglycemia induced ROS-driven metastasis [130, 131]. These observations depict the chemo-preventive role of RSV. In a study conducted by Zheng et al., it was found that between two cell lines of thyroid cancer, one showed susceptibility to RSV treatment, while the other showed resistance [108]. The resistant cell lines encode higher levels of enzymes involved in the metabolism of RSV. These studies highlight the fact that the type of effect a polyphenol will adopt depends on the characteristics and type of malignancy and it is crucial that the treatment approach be tailored accordingly.

6.4 Podophyllotoxin and related compounds

Podophyllotoxin (PT) is a lignan present in the root and rhizome of the herb Mayapple (*Podophyllum peltatum*). It is a cyclo-lignan of the tetra-aryl group that exerts anti-tumor activity by inhibiting microtubule assembly and DNA topoisomerase II [114]. Synthetic anticancer drugs like etiposide and teniposide are used in conventional chemotherapeutic practices, whose structure is based on podophyllotoxin. Etiposide is known to act by inhibiting the activity of DNA-topoisomerase II [120]. Apart from PT itself, several structurally related compounds like picropodophyllotoxin (epimer of PT) and deoxypodophyllotoxin (an analog of PT) have also been found to have anticancer activity in *in vitro* studies (**Table 2**). Synthetic compounds have been designed by modifications of the basic structure of PT or conjugation of PT with other compounds, to derive novel chemical structures with anticancer activity that improves bioavailability, reduces systemic toxic effects, and even counters multi-drug resistance developed by cancer cells [115, 116, 132]. Apart from that, nanomedicines are being designed and synthesized which are modified in a way that facilitates better drug circulation, enhanced drug uptake by tumor cells, and more efficient anti-tumor activity. Two such studies have reported the development of nanomedicines (prodrug loaded nanoparticles), which utilize the oxidatively stressed and acidic environment of cancer cells as cues for site-specific activation and release of pro-drugs, and ultimately induce apoptosis by augmenting the oxidative stress of cancer cells [117, 133]. Liang and Zhou [133] studied the anti-tumor effects of nanoparticles formed from conjugation of homodimers of PT and vitamin K3 into a biocompatible polymer (Pluronic F127). Intracellular metabolism of vitamin K3 increased the oxidative load of cells, which further triggers the release of active PT and acts in synergy to induce cellular apoptosis. In normal cells, due to a lower concentration of ROS, the release of pro-drug is not sufficient to induce cellular death. Xenografted mice model too showed higher anti-tumor activity of the nanodrug due to its higher bioavailability and lesser systemic toxicity [133]. The mechanisms by which PT, its derivatives, and PT-based drugs induce pro-oxidative anticancer effects include induction of oxidative stress and endoplasmic reticulum stress, cell cycle arrest (majorly a G2/M arrest), mitochondrial dysfunction, microtubular damage, modulation of transcriptional factors and signaling pathways like p38, p53, epithelial cadherin, vimentin, snail and slug, inhibition of PI3K-AKT-mTOR pathway, ERK, and induction of oxidative stress-mediated apoptosis (data compiled in **Table 2**).

7. Conclusion

Reactive oxygen species (ROS) can act as both a cause and a cure for cancer. In the present chapter, the prime focus was directed toward discussing the mechanisms by which polyphenolic compounds exert their anti-cancer effects by increasing the oxidative stress level in cancer cells. However, polyphenols can also employ other mechanisms to combat cancer. In a study conducted by Win et al., it was observed that genistein inhibited DNA damage induced by hydrogen peroxide and Cu^{2+}, whereas resveratrol at similar concentration range potentiated the effect and itself caused strand breaks in the presence of metal ions [134]. It indicates that polyphenols can work via different mechanisms to have anti-tumorigenic effects and some polyphenols act as pro-oxidants to deliver the job. Structural characteristics enhancing the antioxidant properties of polyphenols also govern their pro-oxidant roles, which are switched on majorly when metal ions are present or in altered pH conditions. The presence of a metal redox cycle (e.g., Cu^{2+}/Cu^{1+}), formation of polyphenol-metal intermediate complex, or stabilization of intermediate radical are prominent factors that govern the dynamics of the pro-oxidation effect of polyphenols [78]. Orally administered polyphenols undergo extensive metabolism by gut microbial population and phase I and II metabolizing enzymes in the liver and the intestines. The metabolites thus generated may have altered bio-potential. The concentration of polyphenols in systemic circulation is therefore restricted, which may further be subjected to selective absorption in cells due to the presence of membrane efflux pumps. Owing to these factors, it is difficult to achieve a proper effective concentration of polyphenols required to have therapeutic effects, if administered orally. Moreover, the administration of polyphenols in high dosages may lead to the generation of systemic toxicity [135]. Therefore, instead of oral administration, targeted delivery is desirable to manifest anticancer effects. To overcome issues like low bioavailability, reduced circulation time, and improper drug release, nano-sized drug delivery systems (micelles, liposomes, emulsions, metal-, polymer-, and cyclodextrin-based nanoparticles) along with homing ligands and pH/oxidative stress-responsive pro-drug delivery systems are being tested and developed that ensures targeted drug delivery and enhanced anti-tumorigenic response, as already discussed in previous sections. Moreover, the implementation of polyphenol-based chemotherapeutic approaches, either as a mono-therapeutic agent or as an adjunct, is subjective to the cancer type and cellular metabolic profile, as a vast range of heterogeneity exists between different types of cancer. Therefore, extensive *in vivo* and clinical trials are required before the implementation of polyphenol-based anticancer drugs in chemotherapy. This area of research demands a continuous influx of knowledge, ideas, and scientific improvisations which is the need of the hour.

Author details

Aparajita Das* and Sarbani Giri
Molecular and Cell Biology Laboratory, Department of Life Science and Bioinformatics, Assam University, Silchar, Assam, India

*Address all correspondence to: aparajita.das46@gmail.com

References

[1] Phaniendra A, Jestadi DB, Periyasamy L. Free radicals: Properties, sources, targets, and their implication in various diseases. Indian Journal of Clinical Biochemistry. 2015;**30**(1):11-26

[2] Snezhkina AV, Kudryavtseva AV, Kardymon OL, Savvateeva MV, Melnikova NV, Krasnov GS, et al. ROS generation and antioxidant defense systems in normal and malignant cells. Oxidative Medicine and Cellular Longevity. 2019;**2019**:6175804

[3] Onyango AN. Endogenous generation of singlet oxygen and ozone in human and animal tissues: Mechanisms, biological significance, and influence of dietary components. Oxidative Medicine and Cellular Longevity. 2016;**2016**:2398573

[4] El-Bahr SM. Biochemistry of free radicals and oxidative stress. Biochemistry. 2013;**1**(5567/5cjiŋt):11

[5] Andrés CMC, Pérez de la Lastra JM, Juan CA, Plou FJ, Pérez-Lebeña E. Chemistry of hydrogen peroxide formation and elimination in mammalian cells, and its role in various pathologies. Stress. 2022;**2**(3):256-274

[6] Checa J, Aran JM. Reactive oxygen species: Drivers of physiological and pathological processes. Journal of Inflammation Research. 2020;**13**:1057-1073

[7] Bardaweel SK, Gul M, Alzweiri M, Ishaqat A, Ha AL, Bashatwah RM. Reactive oxygen species: The dual role in physiological and pathological conditions of the human body. The Eurasian Journal of Medicine. 2018;**50**(3):193-201

[8] Korczowska-Łącka I, Słowikowski B, Piekut T, Hurła M, Banaszek N, Szymanowicz O, et al. Disorders of endogenous and exogenous antioxidants in neurological diseases. Antioxidants (Basel). 2023;**12**(10):1811

[9] He L, He T, Farrar S, Ji L, Liu T, Ma X. Antioxidants maintain cellular redox homeostasis by elimination of reactive oxygen species. Cellular Physiology and Biochemistry. 2017;**44**(2):532-553

[10] Ding Y, Ye B, Sun Z, Mao Z, Wang W. Reactive oxygen species-mediated pyroptosis with the help of nanotechnology: Prospects for cancer therapy. Advanced NanoBiomed Research. 2022;**3**:2200077

[11] Pandey KB, Rizvi SI. Plant polyphenols as dietary antioxidants in human health and disease. Oxidative Medicine and Cellular Longevity. 2009;**2**(5):270-278

[12] Di Meo S, Reed TT, Venditti P, Victor VM. Role of ROS and RNS sources in physiological and pathological conditions. Oxidative Medicine and Cellular Longevity. 2016;**2016**:1245049

[13] Pizzino G, Irrera N, Cucinotta M, Pallio G, Mannino F, Arcoraci V, et al. Oxidative stress: Harms and benefits for human health. Oxidative Medicine and Cellular Longevity. 2017;**2017**:8416763

[14] Nam TG. Lipid peroxidation and its toxicological implications. Toxicology Research. 2011;**27**(1):1-6

[15] Ayala A, Muñoz MF, Argüelles S. Lipid peroxidation: Production, metabolism, and signaling mechanisms of malondialdehyde and 4-hydroxy-2-nonenal. Oxidative Medicine and Cellular Longevity. 2014;**2014**:360438

[16] Volinsky R, Kinnunen PK. Oxidized phosphatidylcholines in membrane-level

cellular signaling: From biophysics to physiology and molecular pathology. The FEBS Journal. 2013;**280**(12):2806-2816

[17] Yin H, Xu L, Porter NA. Free radical lipid peroxidation: Mechanisms and analysis. Chemical Reviews. 2011;**111**(10):5944-5972

[18] Nimse SB, Pal D. Free radicals, natural antioxidants, and their reaction mechanisms. RSC Advances. 2015;**5**:27986-28006

[19] Cooke MS, Evans MD, Dizdaroglu M, Lunec J. Oxidative DNA damage: Mechanisms, mutation, and disease. The FASEB Journal. 2003;**17**(10):1195-1214

[20] Suzuki YJ. Oxidant-mediated protein amino acid conversion. Antioxidants. 2019;**8**(2):50

[21] Zahra KF, Lefter R, Ali A, Abdellah EC, Trus C, Ciobica A, et al. The involvement of the oxidative stress status in cancer pathology: A double view on the role of the antioxidants. Oxidative Medicine and Cellular Longevity. 2021;**2021**:9965916

[22] Suzuki YJ, Carini M, Butterfield DA. Protein carbonylation. Antioxidants & Redox Signaling. 2010;**12**(3):323-325

[23] Davies MJ. The oxidative environment and protein damage. Biochimica et Biophysica Acta. 2005;**1703**(2):93-109

[24] Kehm R, Baldensperger T, Raupbach J, Höhn A. Protein oxidation - Formation mechanisms, detection and relevance as biomarkers in human diseases. Redox Biology. 2021;**42**:101901

[25] Botelho MC, Teixeira JP, Oliveira PA. Carcinogenesis. In: Wexler P, editor. Encyclopedia of Toxicology. 3rd ed. Oxford, UK: Academic Press; 2014;**1**:713-729

[26] Chaudhary P, Janmeda P, Docea AO, Yeskaliyeva B, Abdull Razis AF, Modu B, et al. Oxidative stress, free radicals and antioxidants: Potential crosstalk in the pathophysiology of human diseases. Frontiers in Chemistry. 2023;**11**:1158198

[27] Khan AQ, Rashid K, AlAmodi AA, Agha MV, Akhtar S, Hakeem I, et al. Reactive oxygen species (ROS) in cancer pathogenesis and therapy: An update on the role of ROS in anticancer action of benzophenanthridine alkaloids. Biomedicine & Pharmacotherapy. 2021;**143**:112142

[28] Iqbal MJ, Kabeer A, Abbas Z, Siddiqui HA, Calina D, Sharifi-Rad J, et al. Interplay of oxidative stress, cellular communication and signaling pathways in cancer. Cell Communication and Signaling: CCS. 2024;**22**(1):7

[29] Son Y, Cheong YK, Kim NH, Chung HT, Kang DG, Pae HO. Mitogen-activated protein kinases and reactive oxygen species: How can ROS activate MAPK pathways? Journal of Signal Transduction. 2011;**2011**:792639

[30] Hayes JD, Dinkova-Kostova AT, Tew KD. Oxidative stress in cancer. Cancer Cell. 2020;**38**(2):167-197

[31] Taguchi K, Motohashi H, Yamamoto M. Molecular mechanisms of the Keap1–Nrf2 pathway in stress response and cancer evolution. Genes to Cells. 2011;**16**(2):123-140

[32] Kwon J, Lee SR, Yang KS, Ahn Y, Kim YJ, Stadtman ER, et al. Reversible oxidation and inactivation of the tumor suppressor PTEN in cells stimulated with peptide growth factors. Proceedings of the National Academy of Sciences

of the United States of America. 2004;**101**(47):16419-16424

[33] Manning BD, Cantley LC. AKT/PKB signaling: Navigating downstream. Cell. 2007;**129**(7):1261-1274

[34] Lei H, Kazlauskas A. A reactive oxygen species-mediated, self-perpetuating loop persistently activates platelet-derived growth factor receptor α. Molecular and Cellular Biology. 2014;**34**(1):110-122

[35] Reuter S, Gupta SC, Chaturvedi MM, Aggarwal BB. Oxidative stress, inflammation, and cancer: How are they linked? Free Radical Biology & Medicine. 2010;**49**(11):1603-1616

[36] Takada Y, Mukhopadhyay A, Kundu GC, Mahabeleshwar GH, Singh S, Aggarwal BB. Hydrogen peroxide activates NF-kappa B through tyrosine phosphorylation of I kappa B alpha and serine phosphorylation of p65: Evidence for the involvement of I kappa B alpha kinase and Syk protein-tyrosine kinase. The Journal of Biological Chemistry. 2003;**278**(26):24233-24241

[37] Labbozzetta M, Notarbartolo M, Poma P. Can NF-κB be considered a valid drug target in neoplastic diseases? Our point of view. International Journal of Molecular Sciences. 2020;**21**:3070

[38] Lee G, Won HS, Lee YM, Choi JW, Oh TI, Jang JH, et al. Oxidative dimerization of PHD2 is responsible for its inactivation and contributes to metabolic reprogramming via HIF-1α activation. Scientific Reports. 2016;**6**:18928

[39] Masoud GN, Li W. HIF-1α pathway: Role, regulation and intervention for cancer therapy. Acta Pharmaceutica Sinica B. 2015;**5**(5):378-389

[40] Song D, Lian Y, Zhang L. The potential of activator protein 1 (AP-1) in cancer targeted therapy. Frontiers in Immunology. 2023;**14**:1224892

[41] Ramundo V, Giribaldi G, Aldieri E. Transforming growth factor-β and oxidative stress in cancer: A crosstalk in driving tumor transformation. Cancers. 2021;**13**(12):3093

[42] Krstić J, Trivanović D, Mojsilović S, Santibanez JF. Transforming growth factor-beta and oxidative stress interplay: Implications in tumorigenesis and cancer progression. Oxidative Medicine and Cellular Longevity. 2015;**2015**:654594

[43] Weinberg F, Ramnath N, Nagrath D. Reactive oxygen species in the tumor microenvironment: An overview. Cancers (Basel). 2019;**11**(8):1191

[44] Arfin S, Jha NK, Jha SK, Kesari KK, Ruokolainen J, Roychoudhury S, et al. Oxidative stress in cancer cell metabolism. Antioxidants (Basel). 2021;**10**(5):642

[45] Toullec A, Gerald D, Despouy G, Bourachot B, Cardon M, Lefort S, et al. Oxidative stress promotes myofibroblast differentiation and tumour spreading. EMBO Molecular Medicine. 2010;**2**(6):211-230

[46] Martinez-Outschoorn UE, Trimmer C, Lin Z, Whitaker-Menezes D, Chiavarina B, Zhou J, et al. Autophagy in cancer associated fibroblasts promotes tumor cell survival: Role of hypoxia, HIF1 induction and NFκB activation in the tumor stromal microenvironment. Cell Cycle. 2010;**9**(17):3515-3533

[47] Liang W, He X, Bi J, Hu T, Sun Y. Role of reactive oxygen species in tumors based on the 'seed and soil' theory: A complex interaction (review). Oncology Reports. 2021;**46**(3):208

[48] Cemerski S, Cantagrel A, Van Meerwijk JP, Romagnoli P. Reactive

oxygen species differentially affect T cell receptor-signaling pathways. The Journal of Biological Chemistry. 2002;**277**(22):19585-19593

[49] Schmielau J, Finn OJ. Activated granulocytes and granulocyte-derived hydrogen peroxide are the underlying mechanism of suppression of T-cell function in advanced cancer patients1. Cancer Research. 2001;**61**(12):4756-4760

[50] Maj T, Wang W, Crespo J, Zhang H, Wang W, Wei S, et al. Oxidative stress controls regulatory T cell apoptosis and suppressor activity and PD-L1-blockade resistance in tumor. Nature Immunology. 2017;**18**(12):1332-1341

[51] Zhang J, Li H, Wu Q, Chen Y, Deng Y, Yang Z, et al. Tumoral NOX4 recruits M2 tumor-associated macrophages via ROS/PI3K signaling-dependent various cytokine production to promote NSCLC growth. Redox Biology. 2019;**22**:101116

[52] Liu Y, Wei J, Guo G, Zhou J. Norepinephrine-induced myeloid-derived suppressor cells block T-cell responses via generation of reactive oxygen species. Immunopharmacology and Immunotoxicology. 2015;**37**(4):359-365

[53] OuYang LY, Wu XJ, Ye SB, Zhang RX, Li ZL, Liao W, et al. Tumor-induced myeloid-derived suppressor cells promote tumor progression through oxidative metabolism in human colorectal cancer. Journal of Translational Medicine. 2015;**13**:47

[54] Kusmartsev S, Gabrilovich DI. Inhibition of myeloid cell differentiation in cancer: The role of reactive oxygen species. Journal of Leukocyte Biology. 2003;**74**(2):186-196

[55] Lin X, Zheng W, Liu J, Zhang Y, Qin H, Wu H, et al. Oxidative stress in malignant melanoma enhances tumor necrosis factor-α secretion of tumor-associated macrophages that promote cancer cell invasion. Antioxidants & Redox Signaling. 2013;**19**(12):1337-1355

[56] Yao Q, Qu X, Yang Q, Wei M, Kong B. CLIC4 mediates TGF-beta1-induced fibroblast-to-myofibroblast transdifferentiation in ovarian cancer. Oncology Reports. 2009;**22**(3):541-548

[57] Ashwin K, Pattanaik AK, Howarth GS. Polyphenolic bioactives as an emerging group of nutraceuticals for promotion of gut health: A review. Food Bioscience. 2021;**44**:101376

[58] Yahfoufi N, Alsadi N, Jambi M, Matar C. The immunomodulatory and anti-inflammatory role of polyphenols. Nutrients. 2018;**10**(11):1618

[59] Zhang Z, Li X, Sang S, McClements DJ, Chen L, Long J, et al. Polyphenols as plant-based nutraceuticals: Health effects, encapsulation, nano-delivery, and application. Food. 2022;**11**(15):2189

[60] Sahiner M, Yilmaz AS, Gungor B, Ayoubi Y, Sahiner N. Therapeutic and nutraceutical effects of polyphenolics from natural sources. Molecules. 2022;**27**(19):6225

[61] Eghbaliferiz S, Iranshahi M. Prooxidant activity of polyphenols, flavonoids, anthocyanins and carotenoids: Updated review of mechanisms and catalyzing metals. Phytotherapy Research. 2016;**30**(9):1379-1391

[62] Rajashekar C. Dual role of plant phenolic compounds as antioxidants and prooxidants. American Journal of Plant Sciences. 2023;**14**:15-28

[63] Yoshino M, Haneda M, Naruse M, Htay HH, Tsubouchi R, Qiao SL, et al.

Prooxidant activity of curcumin: Copper-dependent formation of 8-hydroxy-2′-deoxyguanosine in DNA and induction of apoptotic cell death. Toxicology In Vitro. 2004;**18**(6):783-789

[64] Ahmad A, Syed FA, Singh S, Hadi SM. Prooxidant activity of resveratrol in the presence of copper ions: Mutagenicity in plasmid DNA. Toxicology Letters. 2005;**159**(1):1-12

[65] Azmi AS, Bhat SH, Hadi SM. Resveratrol-Cu(II) induced DNA breakage in human peripheral lymphocytes: Implications for anticancer properties. FEBS Letters. 2005;**579**(14):3131-3135

[66] Farhan M, Zafar A, Chibber S, Khan HY, Arif H, Hadi SM. Mobilization of copper ions in human peripheral lymphocytes by catechins leading to oxidative DNA breakage: A structure activity study. Archives of Biochemistry and Biophysics. 2015;**580**:31-40

[67] Gupte A, Mumper RJ. Elevated copper and oxidative stress in cancer cells as a target for cancer treatment. Cancer Treatment Reviews. 2009;**35**(1):32-46

[68] Gerweck LE, Seetharaman K. Cellular pH gradient in tumor versus normal tissue: Potential exploitation for the treatment of cancer. Cancer Research. 1996;**56**(6):1194-1198

[69] Procházková D, Boušová I, Wilhelmová N. Antioxidant and prooxidant properties of flavonoids. Fitoterapia. 2011;**82**(4):513-523

[70] Yen GC, Duh PD, Tsai HL, Huang SL. Pro-oxidative properties of flavonoids in human lymphocytes. Bioscience, Biotechnology, and Biochemistry. 2003;**67**(6):1215-1222

[71] Zheng LF, Dai F, Zhou B, Yang L, Liu ZL. Prooxidant activity of hydroxycinnamic acids on DNA damage in the presence of Cu(II) ions: Mechanism and structure-activity relationship. Food and Chemical Toxicology. 2008;**46**(1):149-156

[72] Khan NS, Hadi SM. Structural features of tannic acid important for DNA degradation in the presence of Cu(II). Mutagenesis. 1998;**13**(3):271-274

[73] Murakami K, Haneda M, Qiao S, Naruse M, Yoshino M. Prooxidant action of rosmarinic acid: Transition metal-dependent generation of reactive oxygen species. Toxicology In Vitro. 2007;**21**(4):613-617

[74] Labieniec M, Gabryelak T. Study of interactions between phenolic compounds and H2O2 or Cu(II) ions in B14 Chinese hamster cells. Cell Biology International. 2006;**30**(10):761-768

[75] Labieniec M, Gabryelak T. Measurement of DNA damage and protein oxidation after the incubation of B14 Chinese hamster cells with chosen polyphenols. Toxicology Letters. 2005;**155**(1):15-25

[76] Radapong S, Sarker SD, Ritchie KJ. Oxyresveratrol possesses DNA damaging activity. Molecules. 2020;**25**(11):2577

[77] Azmi AS, Bhat SH, Hanif S, Hadi SM. Plant polyphenols mobilize endogenous copper in human peripheral lymphocytes leading to oxidative DNA breakage: A putative mechanism for anticancer properties. FEBS Letters. 2006;**580**(2):533-538

[78] Zheng LF, Wei QY, Cai YJ, Fang JG, Zhou B, Yang L, et al. DNA damage induced by resveratrol and its synthetic analogues in the presence of Cu (II) ions: Mechanism and structure-activity relationship. Free Radical Biology & Medicine. 2006;**41**(12):1807-1816

[79] Dange P, Sarma H, Pandey BN, Mishra KP. Radiation-induced incidence of thymic lymphoma in mice and its prevention by antioxidants. Journal of Environmental Pathology, Toxicology and Oncology. 2007;**26**(4):273-279

[80] Schwarz K, Dobiasch S, Nguyen L, Schilling D, Combs SE. Modification of radiosensitivity by curcumin in human pancreatic cancer cell lines. Scientific Reports. 2020;**10**(1):3815

[81] Hu Y, Cheng L, Du S, Wang K, Liu S. Antioxidant curcumin induces oxidative stress to kill tumor cells (review). Oncology Letters. 2024;**27**(2):67

[82] Urbina-Cano P, Bobadilla-Morales L, Ramírez-Herrera MA, Corona-Rivera JR, Mendoza-Magaña ML, Troyo-Sanromán R, et al. DNA damage in mouse lymphocytes exposed to curcumin and copper. Journal of Applied Genetics. 2006;**47**(4):377-382

[83] Yuan C, Fan R, Zhu K, Wang Y, Xie W, Liang Y. Curcumin induces ferroptosis and apoptosis in osteosarcoma cells by regulating Nrf2/GPX4 signaling pathway. Experimental Biology and Medicine. 2023;**248**(23):2183-2197

[84] Wang C, Song X, Shang M, Zou W, Zhang M, Wei H, et al. Curcumin exerts cytotoxicity dependent on reactive oxygen species accumulation in non-small-cell lung cancer cells. Future Oncology. 2019;**15**(11):1243-1253

[85] Liang WF, Gong YX, Li HF, Sun FL, Li WL, Chen DQ, et al. Curcumin activates ROS signaling to promote pyroptosis in hepatocellular carcinoma HepG2 cells. In Vivo. 2021;**35**(1):249-257

[86] Liu C, Rokavec M, Huang Z, Hermeking H. Curcumin activates a ROS/KEAP1/NRF2/miR-34a/b/c cascade to suppress colorectal cancer metastasis. Cell Death & Differentiation. 2023;**30**(7):1771-1785

[87] Li PM, Li YL, Liu B, Wang WJ, Wang YZ, Li Z. Curcumin inhibits MHCC97H liver cancer cells by activating ROS/TLR-4/caspase signaling pathway. Asian Pacific Journal of Cancer Prevention. 2014;**15**(5):2329-2334

[88] Wu MF, Huang YH, Chiu LY, Cherng SH, Sheu GT, Yang TY. Curcumin induces apoptosis of chemoresistant lung cancer cells via ROS-regulated p38 MAPK phosphorylation. International Journal of Molecular Sciences. 2022;**23**(15):8248

[89] Wang T, Wu X, Al Rudaisat M, Song Y, Cheng H. Curcumin induces G2/M arrest and triggers autophagy, ROS generation and cell senescence in cervical cancer cells. Journal of Cancer. 2020;**11**(22):6704-6715

[90] Cheng Y, Zhao P, Wu S, Yang T, Chen Y, Zhang X, et al. Cisplatin and curcumin co-loaded nano-liposomes for the treatment of hepatocellular carcinoma. International Journal of Pharmaceutics. 2018;**545**(1-2):261-273

[91] Chen M, Qian C, Jin B, Hu C, Zhang L, Wang M, et al. Curcumin analog WZ26 induces ROS and cell death via inhibition of STAT3 in cholangiocarcinoma. Cancer Biology & Therapy. 2023;**24**(1):2162807

[92] Khan AQ, Ahmed EI, Elareer N, Fathima H, Prabhu KS, Siveen KS, et al. Curcumin-mediated apoptotic cell death in papillary thyroid cancer and cancer stem-like cells through targeting of the JAK/STAT3 signaling pathway. International Journal of Molecular Sciences. 2020;**21**(2):438

[93] Wang G, Duan P, Wei Z, Liu F. Curcumin sensitizes carboplatin

treatment in triple negative breast cancer through reactive oxygen species induced DNA repair pathway. Molecular Biology Reports. 2022;**49**(4):3259-3270

[94] Jaganathan SK. Growth inhibition by caffeic acid, one of the phenolic constituents of honey, in HCT 15 colon cancer cells. ScientificWorldJournal. 2012;**2012**:372345

[95] Robati A, Shahsavari Z, Vaezi MA, Safizadeh B, Shirian F, Tavakoli-Yaraki M. Caffeic acid stimulates breast cancer death through reactive oxygen species (ROS) formation, caspase activation and mitochondrial membrane potential depletion. Acta Biochimica Iranica. 2023;**1**(4):176-182

[96] Feriotto G, Tagliati F, Brunello A, Beninati S, Tabolacci C, Mischiati C. A central contribution of TG2 activity to the antiproliferative and pro-apoptotic effects of caffeic acid in K562 cells of human chronic myeloid leukemia. International Journal of Molecular Sciences. 2022;**23**(23):15004

[97] Zhang YX, Yu PF, Gao ZM, Yuan J, Zhang Z. Caffeic acid n-butyl ester-triggered necrosis-like cell death in lung cancer cell line A549 is prompted by ROS mediated alterations in mitochondrial membrane potential. European Review for Medical and Pharmacological Sciences. 2017;**21**(7):1665-1671

[98] Marin EH, Paek H, Li M, Ban Y, Karaga MK, Shashidharamurthy R, et al. Caffeic acid phenethyl ester exerts apoptotic and oxidative stress on human multiple myeloma cells. Investigational New Drugs. 2019;**37**(5):837-848

[99] Lee YY, Kao CL, Tsai PH, Tsai TH, Chiou SH, Wu WF, et al. Caffeic acid phenethyl ester preferentially enhanced radiosensitizing and increased oxidative stress in medulloblastoma cell line. Child's Nervous System. 2008;**24**(9):987-994

[100] Jia J, Yang M, Chen Y, Yuan H, Li J, Cui X, et al. Inducing apoptosis effect of caffeic acid 3,4-dihydroxy-phenethyl ester on the breast cancer cells. Tumour Biology. 2014;**35**(12):11781-11789

[101] Aguilar LE, Jang SR, Park CH, Lee KM. Supramolecular caffeic acid and bortezomib nanomedicine: Prodrug inducing reactive oxygen species and inhibiting cancer cell survival. Pharmaceutics. 2020;**12**(11):1082

[102] Raviadaran R, Ng MH, Chandran D, Ooi KK, Manickam S. Stable W/O/W multiple nanoemulsion encapsulating natural tocotrienols and caffeic acid with cisplatin synergistically treated cancer cell lines (A549 and HEP G2) and reduced toxicity on normal cell line (HEK 293). Materials Science & Engineering C: Materials for Biological Applications. 2021;**121**:111808

[103] Aguilar LE, Chalony C, Kumar D, Park CH, Kim CS. Phenol-boronic surface functionalization of gold nanoparticles; to induce ROS damage while inhibiting the survival mechanisms of cancer cells. International Journal of Pharmaceutics. 2021;**596**:120267

[104] Rodríguez-Enríquez S, Pacheco-Velázquez SC, Marín-Hernández Á, Gallardo-Pérez JC, Robledo-Cadena DX, Hernández-Reséndiz I, et al. Resveratrol inhibits cancer cell proliferation by impairing oxidative phosphorylation and inducing oxidative stress. Toxicology and Applied Pharmacology. 2019;**370**:65-77

[105] Heo JR, Kim SM, Hwang KA, Kang JH, Choi KC. Resveratrol induced reactive oxygen species and endoplasmic reticulum stress-mediated apoptosis, and cell cycle arrest in the A375SM malignant

melanoma cell line. International Journal of Molecular Medicine. 2018;**42**(3):1427-1435

[106] Miki H, Uehara N, Kimura A, Sasaki T, Yuri T, Yoshizawa K, et al. Resveratrol induces apoptosis via ROS-triggered autophagy in human colon cancer cells. International Journal of Oncology. 2012;**40**(4):1020-1028

[107] Kim TH, Park JH, Woo JS. Resveratrol induces cell death through ROS-dependent downregulation of Notch1/PTEN/Akt signaling in ovarian cancer cells. Molecular Medicine Reports. 2019;**19**(4):3353-3360

[108] Zheng X, Jia B, Tian XT, Song X, Wu ML, Kong QY, et al. Correlation of reactive oxygen species levels with resveratrol sensitivities of anaplastic thyroid cancer cells. Oxidative Medicine and Cellular Longevity. 2018;**2018**:6235417

[109] Wang D, Gao Z, Zhang X. Resveratrol induces apoptosis in murine prostate cancer cells via hypoxia-inducible factor 1-alpha (HIF-1α)/reactive oxygen species (ROS)/P53 signaling. Medical Science Monitor. 2018;**24**:8970-8976

[110] Cheng L, Yan B, Chen K, Jiang Z, Zhou C, Cao J, et al. Resveratrol-induced downregulation of NAF-1 enhances the sensitivity of pancreatic cancer cells to gemcitabine via the ROS/Nrf2 signaling pathways. Oxidative Medicine and Cellular Longevity. 2018;**2018**:9482018

[111] Ji S, Zheng Z, Liu S, Ren G, Gao J, Zhang Y, et al. Resveratrol promotes oxidative stress to drive DLC1 mediated cellular senescence in cancer cells. Experimental Cell Research. 2018;**370**(2):292-302

[112] Seino M, Okada M, Shibuya K, Seino S, Suzuki S, Takeda H, et al. Differential contribution of ROS to resveratrol-induced cell death and loss of self-renewal capacity of ovarian cancer stem cells. Anticancer Research. 2015;**35**(1):85-96

[113] Zhang Z, Ji Y, Hu N, Yu Q, Zhang X, Li J, et al. Ferroptosis-induced anticancer effect of resveratrol with a biomimetic nano-delivery system in colorectal cancer treatment. Asian Journal of Pharmaceutical Sciences. 2022;**17**(5):751-766

[114] Lee SO, Joo SH, Kwak AW, Lee MH, Seo JH, Cho SS, et al. Podophyllotoxin induces ROS-mediated apoptosis and cell cycle arrest in human colorectal cancer cells via p38 MAPK signaling. Biomolecules & Therapeutics. 2021;**29**(6):658-666

[115] Choi JY, Cho HJ, Hwang SG, Kim WJ, Kim JI, Um HD, et al. Podophyllotoxin acetate enhances γ-ionizing radiation-induced apoptotic cell death by stimulating the ROS/p38/caspase pathway. Biomedicine & Pharmacotherapy. 2015;**70**:111-118

[116] Li Y, Wang T, Sun Y, Huang T, Li C, Fu Y, et al. p53-mediated PI3K/AKT/mTOR pathway played a role in Ptox(Dpt)-induced EMT inhibition in liver cancer cell lines. Oxidative Medicine and Cellular Longevity. 2019;**2019**:2531493

[117] Li C, Wang Y, Zhang S, Zhang J, Wang F, Sun Y, et al. pH and ROS sequentially responsive podophyllotoxin prodrug micelles with surface charge-switchable and self-amplification drug release for combating multidrug resistance cancer. Drug Delivery. 2021;**28**(1):680-691

[118] Kim KY, Cho HJ, Yu SN, Kim SH, Yu HS, Park YM, et al. Interplay of reactive oxygen species, intracellular

Ca2+ and mitochondrial homeostasis in the apoptosis of prostate cancer cells by deoxypodophyllotoxin. Journal of Cellular Biochemistry. 2013;**114**(5):1124-1134

[119] Kim HS, Oh HN, Kwak AW, Kim E, Lee MH, Seo JH, et al. Deoxypodophyllotoxin inhibits cell growth and induces apoptosis by blocking EGFR and MET in gefitinib-resistant non-small cell lung cancer. Journal of Microbiology and Biotechnology. 2021;**31**(4):559-569

[120] Sang CY, Xu XH, Qin WW, Liu JF, Hui L, Chen SW. DPMA, a deoxypodophyllotoxin derivative, induces apoptosis and anti-angiogenesis in non-small cell lung cancer A549 cells. Bioorganic & Medicinal Chemistry Letters. 2013;**23**(24):6650-6655

[121] Kwak AW, Yoon G, Lee MH, Cho SS, Shim JH, Chae JI. Picropodophyllotoxin, an epimer of podophyllotoxin, causes apoptosis of human esophageal squamous cell carcinoma cells through ROS-mediated JNK/P38 MAPK pathways. International Journal of Molecular Sciences. 2020;**21**(13):4640

[122] Lee SO, Kwak AW, Lee MH, Seo JH, Cho SS, Yoon G, et al. Picropodophyllotoxin induces G1 cell cycle arrest and apoptosis in human colorectal cancer cells via ROS generation and activation of p38 MAPK signaling pathway. Journal of Microbiology and Biotechnology. 2021;**31**(12):1615-1623

[123] Zhu X, Yu Z, Feng L, Deng L, Fang Z, Liu Z, et al. Chitosan-based nanoparticle co-delivery of docetaxel and curcumin ameliorates anti-tumor chemoimmunotherapy in lung cancer. Carbohydrate Polymers. 2021;**268**:118237

[124] Yin D, Hu X, Cai M, Wang K, Peng H, Bai J, et al. Preparation, characterization, and in vitro release of curcumin-loaded IRMOF-10 nanoparticles and investigation of their pro-apoptotic effects on human hepatoma HepG2 cells. Molecules. 2022;**27**(12):3940

[125] Dong Y, Yin S, Song X, Huo Y, Fan L, Ye M, et al. Involvement of ROS-p38-H2AX axis in novel curcumin analogues-induced apoptosis in breast cancer cells. Molecular Carcinogenesis. 2016;**55**(4):323-334

[126] Khan FA, Maalik A, Murtaza G. Inhibitory mechanism against oxidative stress of caffeic acid. Journal of Food and Drug Analysis. 2016;**24**(4):695-702

[127] Maurya DK, Devasagayam TP. Antioxidant and prooxidant nature of hydroxycinnamic acid derivatives ferulic and caffeic acids. Food and Chemical Toxicology. 2010;**48**(12):3369-3373

[128] Li Y, Chen LJ, Jiang F, Yang Y, Wang XX, Zhang Z, et al. Caffeic acid improves cell viability and protects against DNA damage: Involvement of reactive oxygen species and extracellular signal-regulated kinase. Brazilian Journal of Medical and Biological Research. 2015;**48**(6):502-508

[129] Feriotto G, Tagliati F, Giriolo R, Casciano F, Tabolacci C, Beninati S, et al. Caffeic acid enhances the anti-leukemic effect of imatinib on chronic myeloid leukemia cells and triggers apoptosis in cells sensitive and resistant to imatinib. International Journal of Molecular Sciences. 2021;**22**(4):1644

[130] Li W, Cao L, Chen X, Lei J, Ma Q. Resveratrol inhibits hypoxia-driven ROS-induced invasive and migratory ability of pancreatic cancer cells via suppression of the hedgehog signaling pathway. Oncology Reports. 2016;**35**(3):1718-1726

[131] Cao L, Chen X, Xiao X, Ma Q, Li W. Resveratrol inhibits hyperglycemia-driven ROS-induced invasion and migration of pancreatic cancer cells via suppression of the ERK and p38 MAPK signaling pathways. International Journal of Oncology. 2016;**49**(2):735-743

[132] Zhang L, Zhang Z, Chen F, Chen Y, Lin Y, Wang J. Aromatic heterocyclic esters of podophyllotoxin exert anti-MDR activity in human leukemia K562/ADR cells via ROS/MAPK signaling pathways. European Journal of Medicinal Chemistry. 2016;**123**:226-235

[133] Liang B, Zhou D. ROS-activated homodimeric podophyllotoxin nanomedicine with self-accelerating drug release for efficient cancer eradication. Drug Delivery. 2021;**28**(1):2361-2372

[134] Win W, Cao Z, Peng X, Trush MA, Li Y. Different effects of genistein and resveratrol on oxidative DNA damage in vitro. Mutation Research. 2002;**513**(1-2):113-120

[135] Stevenson DE, Hurst RD. Polyphenolic phytochemicals--Just antioxidants or much more? Cellular and Molecular Life Sciences. 2007;**64**(22):2900-2916

Chapter 10

Biomarkers in Aquatic Ecotoxicology: Understanding the Effects of Xenobiotics on the Health of Aquatic Organisms

Mahdi Banaee, Davide Di Paola, Salvatore Cuzzocrea, Marika Cordaro and Caterina Faggio

Abstract

A measurable and/or observable change in a biological or biochemical reaction, encompassing behavioral alterations as well as molecular to physiological levels, is referred to as a biomarker. Biomarker responses must be ecologically meaningful and show exposure to the harmful consequences of environmental stressors. When assessing the condition of an ecosystem, biomarkers are regarded as early warning systems. They are useful for evaluating in-situ chemical exposure and the harmful impacts of contaminants on biota. Although they are assessed on an individual basis, the purpose of their reactions is to forecast population-level consequences. We hope to give a broad definition of biomarkers and xenobiotics in this chapter, as well as an overview of the processes involved in their biotransformation and detoxification in aquatic organisms.

Keywords: ecotoxicology, aquatic organism, biomarkers, pollutant, xenobiotic

1. Introduction

Biomarkers in ecotoxicology have a clear definition. When we talk about biomarkers, it means that we want to define a series of indicators that help to evaluate the toxicity of a chemical compound at different biological levels. Certain biomarkers are used to indicate changes in the physiological status of aquatic animals exposed to environmental pollutants. Some others are useful for monitoring biochemical levels [1, 2]. Moreover, molecular biomarkers reflect responses at cellular and molecular levels in organisms exposed to xenobiotics. Molecular biomarkers include the analysis of gene expression profiles, protein synthesis, or DNA damage in finfish and shellfish exposed to xenobiotics [2].

Changes in growth performance, reproduction success, adaptability, feeding, and behavior in aquatic animals after xenobiotic exposure are physiological biomarkers. Changes in analytes, metabolites, and enzyme activities in tissues and biological fluids such as blood, hemolymph, and cerebrospinal fluid are categorized as biochemical biomarkers. Cellular biomarkers focus on changes in cellular structures, functions, morphology, and membrane integrity. Oxidative biomarkers are used to assay the formation of reactive oxygen species (ROS) and the status of cellular antioxidant defenses [2–5].

We know that biomarkers are reliable indicators for assessing the biological responses of aquatic organisms to xenobiotic exposure. These indicators may provide valuable information about the efficiency of detoxification pathways and the overall health status of aquatic organisms. Therefore, the use of biomarkers for comprehensive evaluation and monitoring of the effects of xenobiotics on aquatic ecosystems is very common. In many cases, the integration of different biomarkers helps researchers to better understand the mechanisms of detoxification and biotransformation [6, 7].

In this chapter, we will first offer a general definition of xenobiotics and outline their biotransformation and detoxification processes. Subsequently, an introduction to biomarkers at various biological levels will be provided.

2. Xenobiotic

Xenobiotics include all materials and compounds that are unknown to living organisms. Most xenobiotics are man-made substances that have various uses. Pharmaceutical materials, agrochemical compounds, crude oil, petroleum derivatives, paint and resin, plastic polymers, industrial and household detergents, cosmetics and health products, etc., are known as xenobiotics (**Figure 1**). Xenobiotics typically enter the environment, particularly aquatic ecosystems, through industrial, urban, and agricultural wastewater, as well as surface runoff, both intentionally and unintentionally. Therefore, their entry into the environment can be considered a serious threat to the life of aquatic organisms and ecosystems. These substances can enter an animal's body through the respiratory tract or gills, skin, and digestive system. Xenobiotic can easily cross biological barriers and enter the blood, cells, and vital tissues due to the hydrophobic and lipophilic nature of these compounds [8].

In response to the entry of xenobiotics into the body of aquatic animals, their detoxification and biotransformation systems activate, attempting to convert these compounds into more polar, simpler, and less toxic substances for elimination from the organism's body. Toxicology studies show that many vital organs, such as the liver, kidneys, digestive system, and respiratory system, can play a significant role in the elimination of metabolites and xenobiotic compounds. Therefore, these organs must have the necessary mechanisms for detoxification and biotransformation of xenobiotic [9–11].

The metabolism of xenobiotic in these organs is often carried out during two stages, including phase I and phase II reactions. In phase I of detoxification, xenobiotic molecules may change their nature through oxidation, reduction, or hydrolysis reactions and become more reactive and water-soluble metabolites. Then these metabolites may enter the second phase reactions for further processing and detoxification. In this phase, the modified xenobiotic compound is combined and conjugated with endogenous molecules, such as glucuronic acid, sulfate, or glutathione. During the conjugation process, the solubility of xenobiotics and their metabolites in water increases and they are easily removed from the animal's body.

Studies showed that xenobiotic can interact with specific receptors on cells, including the aryl hydrocarbon receptor (AhR), pregnane X receptor (PXR),

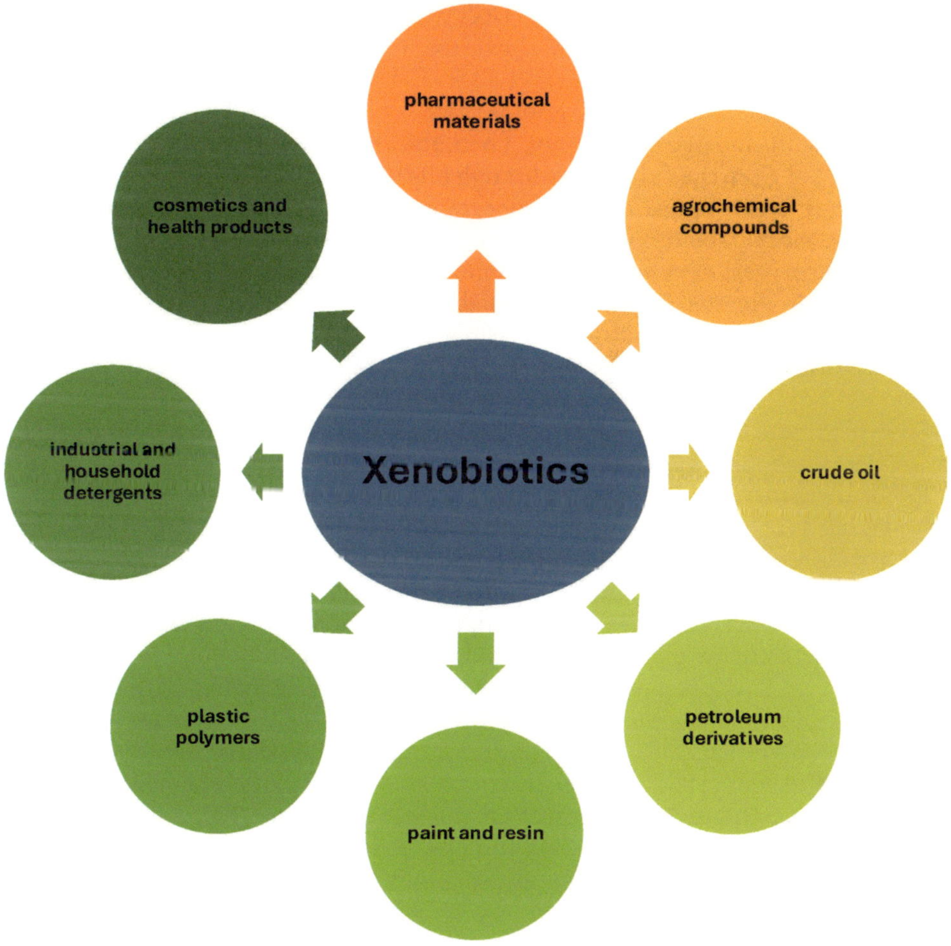

Figure 1.
Most common xenobiotics.

constitutive androstane receptor (CAR), and peroxisome proliferator-activated receptors (PPARs) [12, 13]. The research displayed that the interaction of xenobiotic with cell receptors can cause the activation or inhibition of cell signaling pathways and the occurrence of cell response. In other words, following the binding of xenobiotic to specific receptors, gene expression of enzymes involved in xenobiotic metabolism and detoxification increases. Next, the process of biotransformation, detoxification, and removal of xenobiotic from the aquatic body begins.

The metabolism and biotransformation of xenobiotic is carried out by several enzymatic reactions with the aim of introducing or hiding functional groups that cause xenobiotic reactivity and further change [9–11].

In the second stage of metabolism and biotransformation of xenobiotics, the metabolites produced during the phase I metabolism process are eliminated from the aquatic body through conjugation reactions and combining with polar molecules. Phase II includes glucuronidation, sulfation and combining with glutathione.

In fish liver cells, the glucuronidation process plays an important role in the biotransformation of xenobiotics. During this process, xenobiotics are combined with glucuronic acid, a naturally occurring compound in the body, with the help of the enzyme UDP-glucuronosyltransferase (UGT). During the glucuronidation process, xenobiotics and their metabolites are converted into water-soluble compounds and excreted through urine or bile.

In the sulfation process, sulfate groups can be conjugated with xenobiotics, catalyzed by sulfotransferase enzymes. Then, the resulting sulfate compounds may also be excreted from the fish's body through bile or urine.

The glutathione conjugation is an important biochemical process to remove xenobiotics and their metabolites from the body. Glutathione-S-transferase (GST) can accelerate the process of combining xenobiotics with glutathione. Next, the resulting product is easily excreted from the fish's body through urine and bile [9–11].

2.1 Xenobiotic detoxification pathways

Studies showed that certain enzymes and molecular pathways may be activated during the metabolism and detoxification of xenobiotics in aquatic animals [14–16].

Exposure to different pollutants can induce the Nrf2 pathway. In this situation, Nrf2 may act as an enhancer, and bind to antioxidant response elements (AREs) in the promoter regions of target genes (**Figure 2**).

This phenomenon can affect the expression levels of antioxidant enzymes and phase II detoxification enzymes. Thus, a significant increase in the expression of their genes can mitigate oxidative stress induced by ROS production [16].

Kong et al. studied the mechanisms of Nrf2 and NF-κB signaling pathways in inflammation in *Channa argus* [16].

Studies showed that AhR plays an important role in the cellular uptake of certain xenobiotics. Polycyclic aromatic hydrocarbons (PAHs) and other aromatic compounds may bind to AhR and cross the nuclear membrane. Subsequently, this complex can translocate to the nucleus and bind to response sequence elements. It can then act as either an enhancer or a suppressor, altering the transcriptional process of

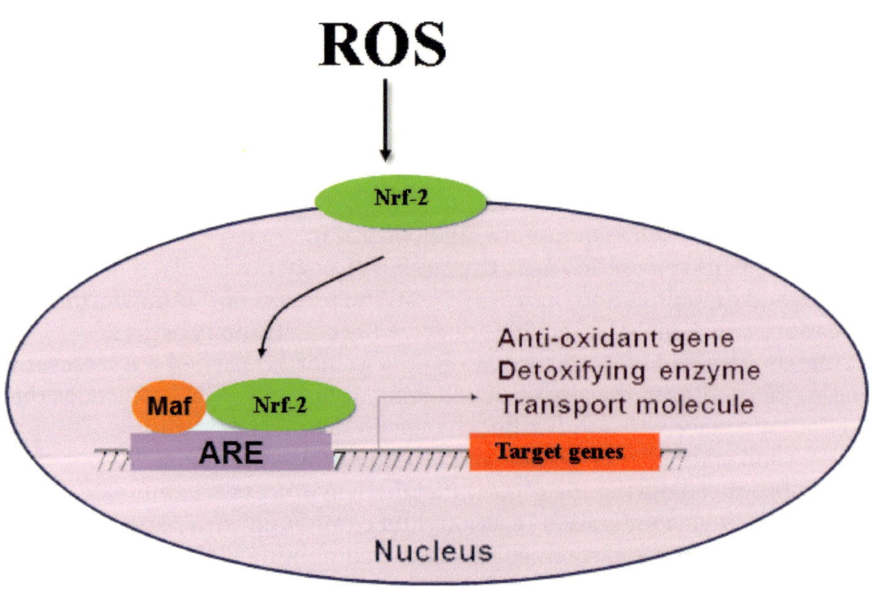

Figure 2.
Nrf2 pathways.

target genes. Studies also showed that activation of the AhR pathway can induce the biosynthesis of cytochrome P450 enzymes, which are involved in the biotransformation of xenobiotics [17].

Similarly, PXR is another nuclear receptor protein that plays a vital role in the biotransformation and detoxification of xenobiotics. PXR is usually found in hepatocytes and intestinal cells. Xenobiotics can bind with PXR and translocate into the nucleus. Once there, this complex binds to response sequence elements in DNA, changing the transcriptional processes of target genes that encode cytochrome P450 enzymes [18, 19].

The function of CAR is similar to other nuclear receptors. CAR is commonly found in hepatocytes. The binding of xenobiotics to response sequence elements is mediated through CAR. The complex of CAR and xenobiotics can influence the expression of genes involved in drug metabolism and detoxification pathways [18, 20].

3. Biomarkers

In toxicology, scientists usually use various parameters to analyze the biological response of organisms to the toxicity of xenobiotics. These biological responses are known as biomarkers. In other words, biomarkers are measurable biological responses

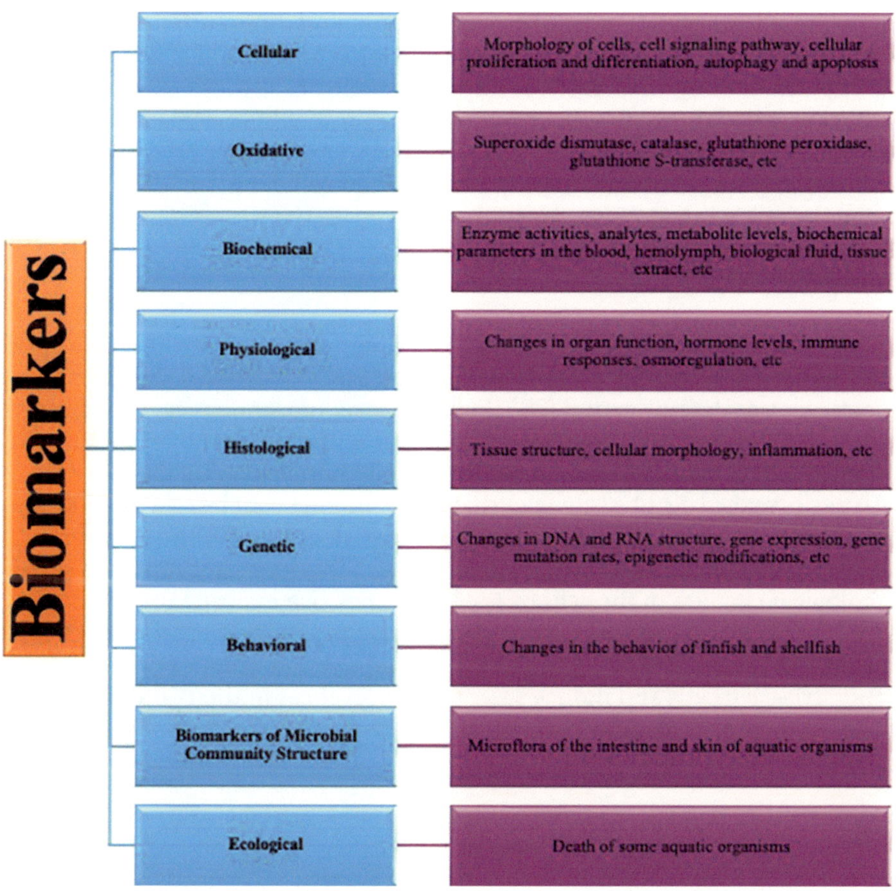

Figure 3.
Different biomarkers that can be measured in aquatic organisms.

of aquatic organisms exposed to environmental pollutants. Ecotoxicologists believe that biomarkers must be sensitive enough to detect changes in aquatic organisms or ecosystems exposed to pollutants. Moreover, the more specific the biomarker, the more accurate the environmental and aquatic health monitoring will be. Additionally, the evaluation and measurement of biomarkers should be practical and feasible, capable of being conducted in both field and laboratory conditions [21, 22].

The biomarkers can be assayed at different levels, including molecular, cellular, tissue, organ, organism, population, and ecosystem levels. Moreover, biomarkers can be used to detect pollutants in the water environment.

Changes in different biomarkers, such as biochemical, physiological, histological, genetic, and behavioral parameters, can help scientists determine exposure periods, toxicity rates of xenobiotics, sensitivity of organisms, risks for aquatic organisms and ecosystems, and more [23–25]. In this part, we will discuss different biomarkers measured in aquatic organisms exposed to environmental contaminants (**Figure 3**).

4. Types of biomarkers

4.1 Cellular biomarkers

Studying cell morphology, cellular proliferation, and differentiation rates, assessing enzyme activities involved in detoxification and biotransformation of xenobiotics, examining cellular stress response, acute phase proteins, cellular signaling pathways, cellular proliferation and apoptosis, and cellular damage and repair biomarkers are important aspects of cellular biomarkers [23, 26, 27]. In this section, we explain the main cellular biomarkers, while other cellular biomarkers are discussed in the following sections.

4.1.1 Morphology of cells

Changes in cellular structure can affect its functions. Changes in the morphology and structure of hepatocytes can disrupt a lot of biochemical and physiological processes, such as detoxification [28]. Furthermore, morphological changes in intestinal, kidney, and gill cells may cause osmoregulation dysfunction. Moreover, morphofunctional changes in the intestine of fish may affect metabolism and growth performance. Studies showed that some anemia may be related to morphological changes in cells of hematopoietic tissues such as the spleen and anterior part of the kidney of fish exposed to xenobiotics [29]. Silva Martinez reported some morphological changes in the cells of the kidneys of fish exposed to municipal sewage. Morphological changes in the gills of *Channa punctata* exposed to heavy metals led to hypoxia [30]. De Castro et al. found that xenobiotics can damage mitochondria-rich cells and induce various interferences in their functions. Some xenobiotics, such as endocrine disruptors, can cause changes in gonadal cellular morphology [31, 32].

4.1.2 Cellular proliferation and differentiation

Studies showed that cellular proliferation in different tissues of fish that are challenged with xenobiotics may be attributed to the regeneration process of damaged tissue [29].

Therefore, an increase in cellular proliferation rate can help tissues to recover themselves after exposure to xenobiotics [33]. However, the collapse in cellular proliferation may lead to tissue necrosis in aquatic animals exposed to different xenobiotics [34, 35].

Studies showed that disruption in stem cell differentiation in the embryos of aquatic animals exposed to xenobiotics may increase the teratogenicity rate in newborns [36, 37]. A lot of examples of teratogenic dysfunction were reported in fish exposed to hospital effluent [38], pesticides [36], and cyanobacteria extract [39].

4.1.3 Cell signaling pathway

Xenobiotics may activate or deactivate different cell signaling pathways. Therefore, understanding the mechanisms of xenobiotics on cell signaling pathways can help scientists mitigate the toxic effects of xenobiotics on aquatic animals. We know that cellular signaling pathways and their cascade of biochemical reactions play an essential role in regulating physiological functions within cells in response to oxidative stress.

Studies have shown that xenobiotics can affect the MAPK (mitogen-activated protein kinase) signaling pathway, inducing cascade biochemical reactions that play important roles in cell growth, differentiation, inflammation, and response to stress [40, 41]. Jia et al. [15] found that hydrogen peroxide could affect the MAPK signaling pathway, changing redox state, apoptosis, and endoplasmic reticulum stress in the hepatocytes of *Cyprinus carpio*. They also showed that the activation of MAPK pathways can induce apoptosis and increase cell death in *C. carpio* exposed to hydrogen peroxide [15].

Moreover, Tian et al. [42] showed that increased levels of ROS in cells can activate MAPK pathways, such as the c-Jun N-terminal kinase (JNK) and p38 MAPK pathways, which are necessary for responding to oxidative stress in *Procambarus clarkii*. The activation of the JNK signaling pathway for the management of oxidative stress, necroptosis, and apoptosis was observed in *C. carpio* exposed to fluoride [43], cadmium [44], and chlorpyrifos [45].

Studies have shown that exposure to xenobiotics can induce the Janus kinase (JAK) and signal transducer and activator of transcription 3 (STAT3) signaling pathway. Upon activation of JAK/STAT3 pathway, STAT proteins are phosphorylated by JAK. Subsequently, they act as transcription factors, regulating the expression of antioxidant enzyme genes. Therefore, activation of the JAK/STAT3 pathway can help manage oxidative stress in fish exposed to xenobiotics [46]. Moreover, Hu et al. [47] found that STAT3 activation can guarantee cell survival in grass carp (*Ctenopharyngodon idella*) by regulating the expression of anti-apoptotic genes and inhibiting pro-apoptotic factors. The role of the JAK/STAT3 pathway in tissue repair and regeneration after oxidative damage is vital and indisputable [48].

4.1.4 Autophagy and apoptosis

Studies showed that autophagy and apoptosis usually occur after inducing signaling pathways of oxidative stress in finfish and shellfish exposed to environmental pollution and toxic materials [49, 50]. Increased autophagy and apoptosis rates were reported in aquatic organisms exposed to different pollutants. The activation of autophagy and apoptosis mechanisms may be a strategy to remove damaged cells following exposure to xenobiotics [51].

4.2 Oxidative biomarkers

Oxidative biomarkers, including antioxidant and non-antioxidant enzymes, ROS contents, and peroxidation metabolites, are used to assess oxidative stress in aquatic organisms exposed to xenobiotics [8, 52]. Superoxide dismutase (SOD), catalase (CAT), glutathione peroxidase (GPx), glutathione S-transferase (GST), etc., are important enzymes involved in neutralizing ROS and protecting cells from oxidative damage (**Figure 4**).

The activity of glutathione reductase (GR) and glucose 6-phosphate dehydrogenase plays crucial roles in regenerating cellular glutathione (GSH) and nicotinamide adenine dinucleotide phosphate (NADPH), respectively, which are necessary for continuing antioxidant enzyme activities [53, 54].

The collapse of cellular antioxidant defense and the imbalance between ROS and antioxidant agents can lead to cellular damage. Therefore, oxidative biomarkers can be used as cellular biomarkers for assessing the toxicity of pollutants in aquatic organisms and ecosystems [55].

Increasing or decreasing levels of antioxidant biomarkers indicate oxidative stress in finfish and shellfish exposed to different environmental pollutants [56]. Changes in oxidative biomarkers were reported in different species of crayfish, such as *Astacus leptodactylus*, *Orconectes limosus*, and *Procambarus clarkii* following exposure to pesticides, polyethylene microplastics, heavy metals, and nano-metals [57–61].

4.3 Biochemical biomarkers

The biochemical biomarkers include measurements of enzyme activities, analytes, metabolite levels, and other biochemical parameters in the blood, hemolymph, biological fluid, and tissue extract. These parameters reflect some physiological

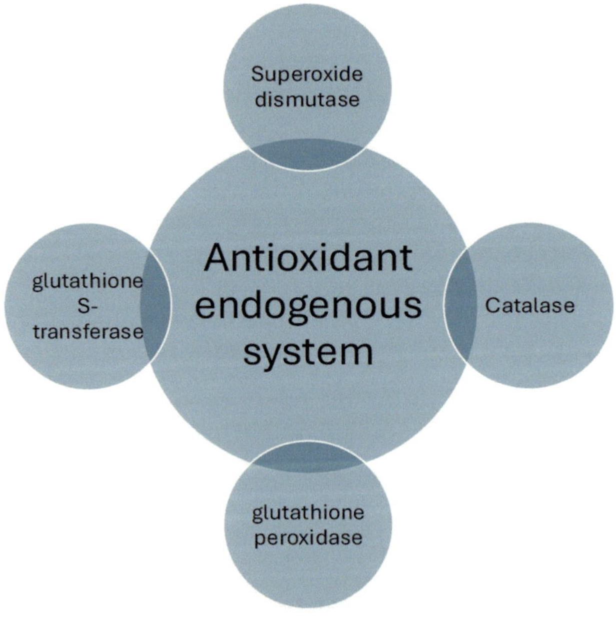

Figure 4.
Some enzymes of the endogenous antioxidant system.

responses to xenobiotic exposure at various levels. For example, the measurement of acetylcholinesterase activity (AChE) in nervous tissues, cerebrospinal fluid, serum, and hemolymph is a common biomarker for exposure to organophosphate and carbamate pesticides, crude oil hydrocarbons, as well as some heavy metals, etc. [62, 63]. A significant decrease in AChE activity was reported in zebrafish (*Danio rerio*), Nile tilapia (*Oreochromis niloticus*), and *Colossoma macropomum* exposed to sulfoxaflor, methidathion, and trichlorfon respectively [64–67]. One of the consequences of oxidative stress is the instability of the cell membrane and the disruption of its physiological function. Therefore, disruption in cellular membrane permeability can lead to the leakage of cytoplasmic enzymes and other analytes and metabolites from damaged cells [68, 69].

The activities of aspartate aminotransferase (AST), alanine aminotransferase (ALT), alkaline phosphatase (ALP), gamma-glutamyl transpeptidase (GGT), and lactate dehydrogenase (LDH) are biomarkers of healthy tissues, especially the liver or hepatopancreas. Therefore, any changes in these levels in the blood or hemolymph suggest tissue damage [70]. Moreover, changes in the lipid profile, total protein, albumin, globulin, glucose, and other analytes in the blood indicate occurrences in the liver, kidney, and gill tissues in aquatic animals exposed to different xenobiotics. Similar changes in blood biochemical parameters were reported in *Channa punctatus* [71], *Cyprinus carpio* [53] and *Alburnus mossulensis* [56] exposed to malathion, diazinon, cadmium chloride, and Bacilar, respectively. The importance of biochemical biomarkers in assessing the health of crayfish exposed to different insecticides, such as thiamethoxam and azoxystrobin, was recently discovered [72, 73]. Rodrigues et al. found that changes in biochemical biomarkers may indicate histopathological damage in the gill tissue of *Lepomis gibbosus* exposed to various xenobiotics [74]. Changes in blood biochemistry and oxidative biomarkers were observed in *Cirrhinus cirrhosis* [75], *Cyprinus carpio* [76], *Mytilus galloprovincialis* [54, 77–79], *Carassius auratus* [80], *Unio tumidus* [81], and *Caridina fossarum* [82] exposed to different xenobiotics.

4.4 Physiological biomarkers

Physiological biomarkers indicate changes in organ function, hormone levels, immune responses, osmoregulation, and other physiological processes that may be affected by pollutants.

In this section, we present some examples of physiological changes reported in different species of aquatic animals exposed to xenobiotics Changes in the various physiological aspects of aquatic animals can assist toxicologists in assessing the toxicity of xenobiotics. For example, exposure to chlorpyrifos significantly affects sex hormone biosynthesis in the gonads of *Oreochromis niloticus* due to oxidative damage [83]. Brander et al. found that some pyrethroid pesticides can act as endocrine disruptors and change the hormonal pathway in fish [84]. Lal showed that reproductive dysfunction in Indian fishes exposed to pesticides may be related to oxidative damage and changes in sex hormone biosynthesis [85]. Changes in thyroid and cortisol hormones were observed in *Cyprinus carpio* exposed to NeemAzal-T/S [86].

Katuli et al. showed that diazinon can affect the osmoregulation mechanism in Caspian roach (*Rutilus rutilus*) fingerlings [87]. Following exposure to xenobiotics, changes in ionic regulation and gill Na+/K + -ATPase activity can reduce the ability of fish to adapt to a new habitat [88]. Osmoregulation dysfunction was reported in *Oncorhynchus mykiss* and *C. carpio*, *Channa punctatus* [89], the African clawed frog [90], and *Litopenaeus vannamei* [91] exposed to different pesticides. Simonato et al.

showed that changes in the ion concentrations in the blood of *Prochilodus lineatus* exposed to diesel oil may indicate osmoregulation dysfunction [92]. Changes in leukocyte counts, cytokine levels, and antibody production, as well as increased sensitivity to pathogens, were observed in aquatic animals exposed to xenobiotics [93]. Changes in various physiological biomarkers were reported in *Mytilus galloprovincialis* exposed to different environmental pollutants [27, 54].

4.5 Histological biomarkers

In histopathology, tissue structure, cellular morphology, inflammation, and any histopathological changes are studied in finfish and shellfish exposed to pollutants. Xenobiotics and their metabolites can induce histopathological damage in vital organs. Therefore, histological studies as biomarkers can help scientists assess the extent of injury in various tissues [94]. Damage to cells, necrosis of tissues, and histopathological damage were observed in the vital organs of aquatic animals exposed to different pollutants [57, 95]. Therefore, histopathological biomarkers are also used as cellular biomarkers in certain situations [96]. Exposure to pesticides can induce histopathological damage to vital tissues of aquatic animals [97, 98]. For example, histopathological changes were reported in the gills, liver, spleen, and kidney of *Cirrhinus mrigala* [99], *Labeo rohita* [100], *Anabas testudineus* [101], *Ctenopharyngodon idella* [102, 103], and *Danio rerio* [104–113] when exposed to different pesticides.

4.6 Genetic biomarkers

Changes in DNA and RNA structure, gene expression, gene mutation rates, and epigenetic modifications are known as genetic biomarkers. The assessment of DNA adduct formation, micronucleus formation, and comet assay indicate DNA damage in finfish and shellfish exposed to xenobiotics. We mentioned that exposure to xenobiotics can induce cell signaling that controls the gene expression of enzymes involved in detoxification. Therefore, the analysis of gene expression of enzymes participating in detoxification can be used as genetic biomarkers for ecotoxicology. Guilherme et al. found that exposure to glyphosate induced DNA damage in *Anguilla Anguilla* [114]. In fact, certain chemicals can bind to DNA and change its structure. Consequently, changes in DNA structure, DNA sequence, and mutations may affect DNA replication and gene expression. Genotoxicity in *Clarias batrachus* was reported after exposure to pendimethalin [115, 116]. DNA damage was reported in *Cnesterodon decemmaculatus* [117], *Oreochromis niloticus* and *Geophagus brasiliensis* [118], *Anguilla Anguilla* [114], and *Prochilodus lineatus* exposed to 3,6-dichloro-2-methoxybenzoic acid, mesotrione, glyphosate, and triclopyr, respectively. Changes in gene expression in the hepatocytes of zebrafish (*Danio rerio*) were observed following exposure to dimethyl phthalate [119]. Derikvandy et al. found that exposure to effluent from the ethyl alcohol industry could result in changes in the gene expression of enzymes involved in the detoxification of xenobiotics in the hepatocytes of *Danio rerio* [120].

4.7 Behavioral biomarkers

Changes in the behavior of finfish and shellfish can be used as behavioral biomarkers. Studies have shown that exposure to certain xenobiotics may disrupt nervous system functions. Therefore, changes in the behavior of aquatic animals may be related to nervous dysfunction [121].

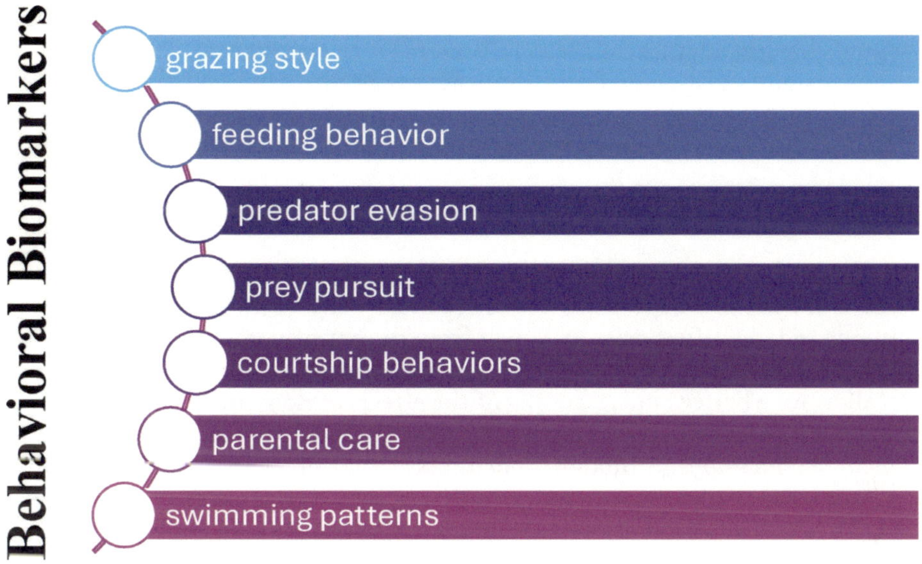

Figure 5.
Behavioral biomarkers.

Changes in feeding behavior, grazing style, predator evasion, prey pursuit, swimming patterns, courtship behaviors, nesting, parental care, territoriality, etc., are important indicators for behavior monitoring. Shiry et al. found that exposure to diazinon led to changes in the behavior and physiology of freshwater swan mussels (**Figure 5**) [122].

Behavioral changes were observed in goldfish [123], rainbow trout [124], and zebrafish exposed to different pollutants [125].

4.8 Biomarkers of microbial community structure

Exposure to different xenobiotics can affect the microflora of the intestine and skin of aquatic organisms. Changes in the variety and frequency of certain microorganisms in the microbiome may impact growth performance, immune response, and overall health of fish and shellfish [126]. Therefore, studying the microbial community of the intestine of aquatic organisms is a suitable biomarker for monitoring the toxicity of xenobiotics. Exposure to pentachlorophenol significantly changed the gut microbial community of *Carassius auratus* [127]. Changes in the intestinal microbiome were observed in *Cyprinus carpio* exposed to trichlorfon, grass carp exposed to cypermethrin, and *Carassius auratus gibelio* exposed to sulfamethoxazole and diazinon [128–130].

4.9 Ecological biomarkers

Exposure to xenobiotics may cause the death of some aquatic organisms within the food web, which can subsequently affect other animals living in the polluted ecosystem [27, 54, 131]. Furthermore, environmental pollutants can affect both biological and non-biological components of water ecosystems. As a consequence, the aquatic ecosystem may attempt to establish a new ecological balance. Therefore, changes in the biological and non-biological components of the ecosystem can be used as ecological biomarkers. Moreover, alterations in the responses of aquatic organisms to ecological changes, such as migration from polluted sites, physiological responses, behavioral changes, may also serve as indicators [122]. Ramya et al. showed that changes in osmoregulation mechanisms in *Mystus keletius* exposed to pesticides could be used as ecological biomarkers to assess environmental health [132].

5. Conclusion

Analytical chemical analysis combined with properly chosen biological endpoints assessed in the tissues of species of concern can help ensure the health of aquatic ecosystems and identify species in danger from the harmful impacts of environmental contaminants. These biological endpoints include the biomarkers that, when combined, allow for a better understanding of the effects of non-chemical stressors while shedding light on questions like contaminant bioavailability, bioaccumulation, and ecological repercussions. Although biomarkers are not a novel concept, their application in ecological risk assessment and natural resource damage assessment has been lacking. These methods can aid in a more thorough evaluation of the intricate consequences of coastal development activities and ecological integrity from a regulatory perspective. In situations where complex mixes of pollutants are prevalent, the use of the mono-biomarker approach is not enough for the assessment of various biological responses that indicate the quality of the environment and for the detection of exposure to contaminants present at low levels in the environment. These pollutants can in fact evolve into irreversible alterations and bring permanent damage to natural populations that are integral parts of aquatic ecosystems. For this reason, researchers contribute to the development of a multi-biomarker approach, able to investigate the different toxicological responses in aquatic organisms, which represents a fundamental tool for more accurate environmental monitoring of the population and the potential danger deriving from contamination [133]. It is necessary to evaluate the clinical risk associated with the detected hazard, which is contingent upon several parameters, such as patient health, metabolic stability, pharmacodynamic response, exposure, and indication. While identifying hazards is comparatively simple, assessing risks is a complex and demanding process.

Author details

Mahdi Banaee[1†], Davide Di Paola[2†], Salvatore Cuzzocrea[2], Marika Cordaro[3*] and Caterina Faggio[2,4]

1 Faculty of Natural Resources and the Environment, Aquaculture Department, Behbahan Khatam Alanbia University of Technology, Behbahan, Iran

2 Department of Chemical, Biological, Pharmaceutical and Environmental Sciences, University of Messina, Messina, Italy

3 Department of Biomedical and Dental Sciences and Morphofunctional Imaging, University of Messina, Messina, Italy

4 Department of Eco-Sustainable Marine Biotechnology, Stazione Zoologica Anton Dohrn, Naples, Italy

*Address all correspondence to: cordarom@unime.it

† These authors equally contributed to this work.

References

[1] Stara A et al. Effects of long-term exposure of *Mytilus galloprovincialis* to thiacloprid: A multibiomarker approach. Environmental Pollution. 2021;**289**:117892

[2] Saha S et al. Longer-term adverse effects of selenate exposures on hematological and serum biochemical variables in air-breathing fish *Channa punctata* (Bloch, 1973) and non-air breathing fish *Ctenopharyngodon Idella* (Cuvier, 1844): An integrated biomarker response approach. Biological Trace Element Research. 2023;**201**(7):3497-3512

[3] Abdel-Tawwab M et al. Fish response to hypoxia stress: Growth, physiological, and immunological biomarkers. Fish Physiology and Biochemistry. 2019;**45**(3):997-1013

[4] Petrovici A et al. Toxicity of deltamethrin to zebrafish gonads revealed by cellular biomarkers. Journal of Marine Science and Engineering. 2020;**8**(2):73

[5] Dhara K et al. Fluoride sensitivity in freshwater snail, Bellamya bengalensis (Lamarck, 1882): An integrative biomarker response assessment of behavioral indices, oxygen consumption, haemocyte and tissue protein levels under environmentally relevant exposure concentrations. Environmental Toxicology and Pharmacology. 2022;**89**:103789

[6] Burgos-Aceves MA et al. Ecotoxicological perspectives of microplastic pollution in amphibians. Journal of Toxicology and Environmental Health. Part B, Critical Reviews. 2022;**25**(8):405-421

[7] Multisanti CR et al. Sentinel species selection for monitoring microplastic pollution: A review on one health approach. Ecological Indicators. 2022;**145**:109587

[8] Banaee M et al. Evaluation of single and combined effects of cadmium and micro-plastic particles on biochemical and immunological parameters of common carp (*Cyprinus carpio*). Chemosphere. 2019;**236**:124335

[9] Mahdi B. Adverse effect of insecticides on various aspects of fish's biology and physiology. In: SS, Marcelo L, editors. Insecticides. IntechOpen: Rijeka; 2012. p. Ch. 6

[10] Banaee M. Physiological Dysfunction in fish after insecticides exposure. Insecticides - Development of Safer and More Effective Technologies. 2013. pp. 103e143

[11] Mahdi B. Toxicological interaction effects of herbicides and the environmental pollutants on aquatic organisms. In: Kassio Ferreira M, editor. New Insights in Herbicide Science. IntechOpen: Rijeka; 2022. p. Ch. 6

[12] King-Heiden TC et al. Reproductive and developmental toxicity of dioxin in fish. Molecular and Cellular Endocrinology. 2012;**354**(1-2):121-138

[13] Bainy AC et al. Functional characterization of a full length pregnane X receptor, expression in vivo, and identification of PXR alleles, in zebrafish (*Danio rerio*). Aquatic Toxicology. 2013;**142-143**:447-457

[14] Chen X et al. Identification of the Nrf2 in the fathead minnow muscle cell line: Role for a regulation in response to H(2)O(2) induced the oxidative stress in fish cell. Fish Physiology and Biochemistry. 2020;**46**(5):1699-1711

[15] Jia R et al. Chronic exposure of hydrogen peroxide alters redox state, apoptosis and endoplasmic reticulum stress in common carp (*Cyprinus carpio*). Aquatic Toxicology. 2020;**229**:105657

[16] Kong Y et al. Effects of dietary curcumin inhibit deltamethrin-induced oxidative stress, inflammation and cell apoptosis in *Channa argus* via Nrf2 and NF-κB signaling pathways. Aquaculture. 2021;**540**:736744

[17] Marx-Stoelting P, Knebel C, Braeuning A. The connection of azole fungicides with xeno-sensing nuclear receptors, drug metabolism and hepatotoxicity. Cells. 2020;**9**(5):1-25

[18] Gao J et al. Hepatic expression patterns of aryl hydrocarbon receptor, pregnane X receptor, two cytochrome P450s and five phase II metabolism genes responsive to 17alpha-methyltestosterone in rare minnow *Gobiocypris rarus*. Environmental Toxicology and Pharmacology. 2014;**37**(3):1157-1168

[19] Lange A et al. Development of a common carp (*Cyprinus carpio*) pregnane X receptor (cPXR) transactivation reporter assay and its activation by azole fungicides and pharmaceutical chemicals. Toxicology In Vitro. 2017;**41**:114-122

[20] Rooney JP et al. Chemical activation of the constitutive androstane receptor leads to activation of oxidant-induced Nrf2. Toxicological Sciences. 2019;**167**(1):172-189

[21] Banaee M, Mohammadipour S, Madhani S. Effects of sublethal concentrations of permethrin on bioaccumulation of cadmium in zebra cichlid (*Cichlasoma nigrofasciatum*). Toxicological and Environmental Chemistry. 2015;**97**:1-8

[22] Banaee M et al. Assessing metal toxicity on crustaceans in aquatic ecosystems: A comprehensive review. Biological Trace Element Research. 2024:1-19

[23] Ahmadi K et al. Effects of long-term diazinon exposure on some immunological and haematological parameters in rainbow trout *Oncorhynchus mykiss* (Walbaum, 1792). Toxicology and Environmental Health Sciences. 2014;**6**:1-7

[24] Banaee M et al. Effects of microplastic exposure on the blood biochemical parameters in the pond turtle (*Emys orbicularis*). Environmental Science and Pollution Research International. 2021;**28**(8):9221-9234

[25] Banihashemi EA et al. Effect of microplastics on Yersinia ruckeri infection in rainbow trout (*Oncorhynchus mykiss*). Environmental Science and Pollution Research International. 2022;**29**(8):11939-11950

[26] Fabrello J et al. Identification of haemocytes and histological examination of gills of the spiny oyster *Spondylus gaederopus* (Linnaeus, 1758). Fish and Shellfish Immunology. 2022;**130**:164-174

[27] Impellitteri F et al. The odd couple: Caffeine and microplastics. Morphological and physiological changes in *Mytilus galloprovincialis*. Microscopy Research and Technique. 2024;**87**(5):1092-1110

[28] Rodd AL et al. A 3D fish liver model for aquatic toxicology: Morphological changes and Cyp1a induction in PLHC-1 microtissues after repeated benzo(a) pyrene exposures. Aquatic Toxicology. 2017;**186**:134-144

[29] Witeska M, Kondera E, Bojarski B. Hematological and hematopoietic

analysis in fish toxicology—A review. Animals. 2023;**13**(16):2625

[30] Pandey S et al. Effects of exposure to multiple trace metals on biochemical, histological and ultrastructural features of gills of a freshwater fish, *Channa punctata Bloch*. Chemico-Biological Interactions. 2008;**174**(3):183-192

[31] Sachi ITC et al. Biochemical and morphological biomarker responses in the gills of a Neotropical fish exposed to a new flavonoid metal-insecticide. Ecotoxicology and Environmental Safety. 2021;**208**:111459

[32] Carnevali O et al. Endocrine-disrupting chemicals in aquatic environment: What are the risks for fish gametes? Fish Physiology and Biochemistry. 2018;**44**(6):1561-1576

[33] Sales CF et al. Proliferation, survival and cell death in fish gills remodeling: From injury to recovery. Fish and Shellfish Immunology. 2017;**68**:10-18

[34] Costa PM et al. Alterations to proteome and tissue recovery responses in fish liver caused by a short-term combination treatment with cadmium and benzo[a]pyrene. Environmental Pollution. 2010;**158**(10):3338-3346

[35] Segner H et al. Immunotoxicity of xenobiotics in fish: A role for the aryl hydrocarbon receptor (AhR)? International Journal of Molecular Sciences. 2021;**22**(17):1-27

[36] Pašková V, Hilscherová K, Bláha L. Teratogenicity and embryotoxicity in aquatic organisms after pesticide exposure and the role of oxidative stress. Reviews of Environmental Contamination and Toxicology. 2011;**211**:25-61

[37] Garcia-Käufer M et al. Genotoxic and teratogenic effect of freshwater sediment samples from the Rhine and Elbe River (Germany) in zebrafish embryo using a multi-endpoint testing strategy. Environmental Science and Pollution Research International. 2015;**22**(21):16341-16357

[38] Luja-Mondragón M et al. Alterations to embryonic development and teratogenic effects induced by a hospital effluent on *Cyprinus carpio* oocytes. Science of the Total Environment. 2019;**660**:751-764

[39] Jonas A et al. Endocrine, teratogenic and neurotoxic effects of cyanobacteria detected by cellular in vitro and zebrafish embryos assays. Chemosphere. 2015;**120**:321-327

[40] Jing H et al. Pb exposure triggers MAPK-dependent inflammation by activating oxidative stress and miRNA-155 expression in carp head kidney. Fish and Shellfish Immunology. 2020;**106**:219-227

[41] Zhu R et al. Unlocking the potential of N-acetylcysteine: Improving hepatopancreas inflammation, antioxidant capacity and health in common carp (*Cyprinus carpio*) via the MAPK/NF-κB/Nrf2 signalling pathway. Fish and Shellfish Immunology. 2024;**144**:109294

[42] Tian H et al. c-Jun N-terminal kinase (JNK) in *Procambarus clarkii*: Molecular characterization and involvement in oxidative stress-induced apoptosis during molting cycle. Comparative Biochemistry and Physiology. Part B, Biochemistry and Molecular Biology. 2022;**257**:110676

[43] Cao J et al. Protective properties of sesamin against fluoride-induced oxidative stress and apoptosis in kidney of carp (*Cyprinus carpio*) via JNK signaling pathway. Aquatic Toxicology. 2015;**167**:180-190

[44] Zhang J et al. Cadmium-induced oxidative stress promotes apoptosis and necrosis through the regulation of the miR-216a-PI3K/AKT axis in common carp lymphocytes and antagonized by selenium. Chemosphere. 2020;**258**:127341

[45] Chen J et al. Chlorpyrifos caused necroptosis via MAPK/NF-κB/TNF-α pathway in common carp (*Cyprinus carpio L.*) gills. Comparative Biochemistry and Physiology Part C: Toxicology and Pharmacology. 2021;**249**:109126

[46] Wang D et al. Koumine induces apoptosis in *Cyprinus carpio* liver cells by regulating JAK-STAT and p53 signaling pathways. Fish and Shellfish Immunology. 2023;**132**:108475

[47] Hu Q-Y et al. Antimicrobial peptide Isalo scorpion cytotoxic peptide (IsCT) enhanced growth performance and improved intestinal immune function associated with janus kinases (JAKs)/signal transducers and activators of transcription (STATs) signalling pathways in on-growing grass carp (*Ctenopharyngodon idella*). Aquaculture. 2021;**539**:736585

[48] Elsaeidi F et al. Jak/Stat signaling stimulates zebrafish optic nerve regeneration and overcomes the inhibitory actions of Socs3 and Sfpq. The Journal of Neuroscience. 2014;**34**(7):2632-2644

[49] AnvariFar H et al. Environmental pollution and toxic substances: Cellular apoptosis as a key parameter in a sensible model like fish. Aquatic Toxicology. 2018;**204**:144-159

[50] Silvestre F. Signaling pathways of oxidative stress in aquatic organisms exposed to xenobiotics. Journal of Experimental Zoology Part A: Ecological and Integrative Physiology. 2020;**333**(6):436-448

[51] AnvariFar H et al. Apoptosis in fish: Environmental factors and programmed cell death. Cell and Tissue Research. 2017;**368**(3):425-439

[52] Ibrahim ATA, Banaee M, Sureda A. Genotoxicity, oxidative stress, and biochemical biomarkers of exposure to green synthesized cadmium nanoparticles in *Oreochromis niloticus (L.)*. Comparative Biochemistry and Physiology Part C: Toxicology and Pharmacology. 2021;**242**:108942

[53] Banaei M, Forouzanfar M, Jafarinia M. Toxic effects of polyethylene microplastics on transcriptional changes, biochemical response, and oxidative stress in common carp (*Cyprinus carpio*). Comparative Biochemistry and Physiology Part C: Toxicology and Pharmacology. 2022;**261**:109423

[54] Impellitteri F et al. Chlorpromazine's impact on *Mytilus galloprovincialis*: A multi-faceted investigation. Chemosphere. 2023;**350**:141079

[55] Gokul T et al. Impact of particulate pollution on aquatic invertebrates. Environmental Toxicology and Pharmacology. 2023;**100**:104146

[56] Banaee M et al. Effects of cadmium chloride and biofertilizer (Bacilar) on biochemical parameters of freshwater fish, *Alburnus mossulensis*. Comparative Biochemistry and Physiology Part C: Toxicology and Pharmacology. 2023;**268**:109614

[57] Gao X et al. Effect of maduramicin on crayfish (*Procambius clarkii*): Hematological parameters, oxidative stress, histopathological changes and stress response. Ecotoxicology and Environmental Safety. 2021;**211**:111896

[58] Gholamhosseini A et al. Physiological response of freshwater crayfish, *Astacus leptodactylus* exposed to polyethylene microplastics at different temperature. Comparative Biochemistry and Physiology Part C: Toxicology and Pharmacology. 2023;**267**:109581

[59] Zeidi M, Park CB, Il Kim C. Synergism effect between nanofibrillation and interface tuning on the stiffness-toughness balance of rubber-toughened polymer nanocomposites: A multiscale analysis. ACS Applied Materials and Interfaces. 2023;**15**(20):24948-24967

[60] Zhang Y et al. Cadmium-induced oxidative stress, histopathology, and transcriptome changes in the hepatopancreas of freshwater crayfish (*Procambarus clarkii*). Science of the Total Environment. 2019;**666**:944-955

[61] Struzynski W et al. Markers of oxidative stress in hepatopancreas of crayfish (*Orconectes limosus*, raf) experimentally exposed to nanosilver. Environmental Toxicology. 2014;**29**(11):1283-1291

[62] Bernal-Rey DL et al. Seasonal variations in the dose-response relationship of acetylcholinesterase activity in freshwater fish exposed to chlorpyrifos and glyphosate. Ecotoxicology and Environmental Safety. 2020;**187**:109673

[63] Olivares-Rubio HF, Espinosa-Aguirre JJ. Acetylcholinesterase activity in fish species exposed to crude oil hydrocarbons: A review and new perspectives. Chemosphere. 2021;**264** (Pt 1):128401

[64] Piner Benli P, Çelik M. Glutathione and its dependent enzymes' modulatory responses to neonicotinoid insecticide sulfoxaflor induced oxidative damage in zebrafish in vivo. Science Progress. 2021;**104**(2):00368504211028361

[65] Liu Y et al. Acetylcholinesterase activity monitoring and natural anti-neurological disease drug screening via rational design of deep eutectic solvents and CeO(2)-Co(OH)(2) nanosheets. Analytical Chemistry. 2022;**94**(15):5970-5979

[66] Liu YH et al. Acetylcholinesterase inhibitory activity and neuroprotection in vitro, molecular docking, and improved learning and memory functions of demethylcurcumin in scopolamine-induced amnesia ICR mice. Food and Function. 2020;**11**(3):2328-2338

[67] Duncan WP et al. Acute toxicity of the pesticide trichlorfon and inhibition of acetylcholinesterase in *Colossoma macropomum* (Characiformes: Serrasalmidae). Aquaculture International. 2020;**28**:815-830

[68] Banaee M et al. Effects of diazinon on biochemical parameters of blood in rainbow trout (*Oncorhynchus mykiss*). Pesticide Biochemistry and Physiology. 2011;**99**(1):1-6

[69] Rohani MF. Pesticides toxicity in fish: Histopathological and hemato-biochemical aspects–A review. Emerging Contaminants. 2023;**9**(3):100234

[70] Banaee N. Enhanced dose measurement of zinc oxide nanoparticles by radiochromic polymer dosimeter and Monte Carlo simulation. Reports of Practical Oncology and Radiotherapy. 2020;**25**(4):515-520

[71] Bharti S, Rasool F. Analysis of the biochemical and histopathological impact of a mild dose of commercial malathion on *Channa punctatus* (Bloch) fish. Toxicology Reports. 2021;**8**:443-455

[72] Uckun M, Ozmen M. Evaluating multiple biochemical markers in *Xenopus laevis* tadpoles exposed to the pesticides thiacloprid and trifloxystrobin in single and mixed forms. Environmental Toxicology and Chemistry. 2021;**40**(10):2846-2860

[73] Uckun AA, Oz OB. Evaluation of the acute toxic effect of azoxystrobin on non-target crayfish (*Astacus leptodactylus* Eschscholtz, 1823) by using oxidative stress enzymes, ATPases and cholinesterase as biomarkers. Drug and Chemical Toxicology. 2021;**44**(5):550-557

[74] Rodrigues S et al. Alterations in gills of *Lepomis gibbosus*, after acute exposure to several xenobiotics (pesticide, detergent and pharmaceuticals): Morphometric and biochemical evaluation. Drug and Chemical Toxicology. 2015;**38**(2):126-132

[75] Hamidi M et al. Evaluation of two fungal exopolysaccharides as potential biomaterials for wound healing applications. World Journal of Microbiology and Biotechnology. 2022;**39**(2):49

[76] Machanlou M et al. Study on the hematological toxicity of *Cyprinus carpio* exposed to water-soluble fraction of crude oil and TiO(2) nanoparticles in the dark and ultraviolet. Chemosphere. 2023;**343**:140272

[77] Impellitteri F et al. Physiological and biochemical responses to caffeine and microplastics in *Mytilus galloprovincialis*. Science of the Total Environment. 2023;**890**:164075

[78] Impellitteri F et al. Cellular and oxidative stress responses of *Mytilus galloprovincialis* to chlorpromazine: Implications of an antipsychotic drug exposure study. Frontiers in Physiology. 2023;**14**:1267953

[79] Impellitteri F et al. Exploring the impact of contaminants of emerging concern on fish and invertebrates physiology in the Mediterranean Sea. Biology (Basel). 2023;**12**(6):767

[80] Sadeghi S et al. Copper-oxide nanoparticles effects on goldfish (*Carassius auratus*): Lethal toxicity, haematological, and biochemical effects. Veterinary Research Communications. 2024;**48**:1611-1620

[81] Martyniuk V et al. Combined effect of microplastic, salinomycin and heating on *Unio tumidus*. Environmental Toxicology and Pharmacology. 2023;**98**:104068

[82] Gholamhosseini A et al. Individual and combined impact of microplastics and lead acetate on the freshwater shrimp (*Caridina fossarum*): Biochemical effects and physiological responses. Journal of Contaminant Hydrology. 2024;**262**:104325

[83] Oruc E. Effects of diazinon on antioxidant defense system and lipid peroxidation in the liver of *Cyprinus carpio (L.)*. Environmental Toxicology. 2011;**26**(6):571-578

[84] Brander SM et al. Pyrethroid pesticides as endocrine disruptors: Molecular mechanisms in vertebrates with a focus on fishes. Environmental Science and Technology. 2016;**50**(17):8977-8992

[85] Lal B. Pesticide—Induced reproductive dysfunction in Indian fishes. Fish Physiology and Biochemistry. 2007;**33**:455-462

[86] Korkmaz N, Örün İ. Effects of pesticide neemazal-T/S on thyroid, stress hormone and some cytokines levels in freshwater common carp, *Cyprinus carpio L.* Toxin Reviews. 2022;**41**(2):496-505

[87] Katuli KK et al. Impact of a short-term diazinon exposure on the osmoregulation potentiality of Caspian roach (*Rutilus rutilus*) fingerlings. Chemosphere. 2014;**108**:396-404

[88] Suvetha L, Ramesh M, Saravanan M. Influence of cypermethrin toxicity on ionic regulation and gill Na(+)/K(+)-ATPase activity of a freshwater teleost fish *Cyprinus carpio*. Environmental Toxicology and Pharmacology. 2010;**29**(1):44-49

[89] Agrahari S, Gopal K. Inhibition of Na+–K+-ATPase in different tissues of freshwater fish *Channa punctatus* (Bloch) exposed to monocrotophos. Pesticide Biochemistry and Physiology. 2008;**92**(2):57-60

[90] Alvarez-Vergara F, Sanchez-Hernandez JC, Sabat P. Biochemical and osmoregulatory responses of the African clawed frog experimentally exposed to salt and pesticide. Comparative Biochemistry and Physiology Part C: Toxicology and Pharmacology. 2022;**258**:109367

[91] Galindo-Reyes JG et al. Enzymatic and osmoregulative alterations in white shrimp *Litopenaeus vannamei* exposed to pesticides. Chemosphere. 2000;**40**(3):233-237

[92] Simonato JD, Guedes CL, Martinez CB. Biochemical, physiological, and histological changes in the neotropical fish *Prochilodus lineatus* exposed to diesel oil. Ecotoxicology and Environmental Safety. 2008;**69**(1):112-120

[93] Yang C, Lim W, Song G. Immunotoxicological effects of insecticides in exposed fishes. Comparative Biochemistry and Physiology Part C: Toxicology and Pharmacology. 2021;**247**:109064

[94] Wolf JC, Wheeler JR. A critical review of histopathological findings associated with endocrine and non-endocrine hepatic toxicity in fish models. Aquatic Toxicology. 2018;**197**:60-78

[95] Guerrera MC et al. Micro and nano plastics distribution in fish as model organisms: Histopathology, blood response and bioaccumulation in different organs. Applied Sciences. 2021;**11**(13):5768

[96] Gholamhosseini A et al. Molecular, histopathologic and electron microscopic analysis of white spot syndrome virus in wild shrimp (*Fenneropenaeus indicus*) in the coastal waters of Iran. Archives of Virology. 2020;**165**(6):1433-1440

[97] Banaee M et al. Histopathological alterations induced by diazinon in rainbow trout (*Oncorhynchus mykiss*). International Journal of Environmental Research. 2013;7:735-744

[98] Banaee M et al. Biochemical and histological changes in the liver tissue of rainbow trout (*Oncorhynchus mykiss*) exposed to sub-lethal concentrations of diazinon. Fish Physiology and Biochemistry. 2013;**39**(3):489-501

[99] Ghayyur S et al. A comparative study on the effects of selected pesticides on hemato-biochemistry and tissue histology of freshwater fish *Cirrhinus mrigala* (Hamilton, 1822). Saudi Journal of Biological Sciences. 2021;**28**(1):603-611

[100] Nataraj B et al. Hepatic oxidative stress, genotoxicity and histopathological alteration in fresh water fish *Labeo rohita* exposed to organophosphorus pesticide profenofos. Biocatalysis and Agricultural Biotechnology. 2017;**12**:185-190

[101] Velmurugan B et al. Cytological and histological effects of pesticide chlorpyriphos in the gills of *Anabas*

testudineus. Drug and Chemical Toxicology. 2020;**43**(4):409-414

[102] Vajargah MF et al. Histological effects of sublethal concentrations of insecticide lindane on intestinal tissue of grass carp (*Ctenopharyngodon idella*). Veterinary Research Communications. 2021;**45**(4):373-380

[103] Azadikhah D et al. Hematological and histopathological changes of juvenile grass carp (*Ctenopharyngodon idella*) exposed to lethal and sublethal concentrations of roundup (glyphosate 41% SL). Aquaculture Research. 2023;(1):4351307

[104] D'Iglio C et al. Toxic effects of gemcitabine and paclitaxel combination: Chemotherapy drugs exposure in zebrafish. Toxics. 2023;**11**(6):554

[105] Di Paola D et al. Early exposure to environmental pollutants: Imidacloprid potentiates cadmium toxicity on zebrafish retinal cells death. Animals (Basel). 2022;**12**(24):3484

[106] Di Paola D et al. Impact of mycotoxin contaminations on aquatic organisms: Toxic effect of aflatoxin B1 and Fumonisin B1 mixture. Toxins (Basel). 2022;**14**(8)

[107] Di Paola D et al. Environmental risk assessment of dexamethasone sodium phosphate and tocilizumab mixture in zebrafish early life stage (*Danio rerio*). Toxics. 2022;**10**(6):279

[108] Di Paola D et al. Environmental co-exposure to potassium perchlorate and Cd caused toxicity and thyroid endocrine disruption in zebrafish embryos and larvae (*Danio rerio*). Toxics. 2022;**10**(4):198

[109] Di Paola D et al. Environmental risk assessment of oxaliplatin exposure on early life stages of zebrafish (*Danio rerio*). Toxics. 2022;**10**(2):81

[110] Di Paola D et al. Assessment of 2-Pentadecyl-2-oxazoline role on lipopolysaccharide-induced inflammation on early stage development of zebrafish (*Danio rerio*). Life (Basel). 2022;**12**(1):128

[111] Di Paola D et al. Intestinal disorder in zebrafish larvae (*Danio rerio*): The protective action of N-Palmitoylethanolamide-oxazoline. Life (Basel). 2022;**12**(1):125

[112] Di Paola D et al. Combined toxicity of Xenobiotics Bisphenol a and heavy metals on zebrafish embryos (*Danio rerio*). Toxics. 2021;**9**(12):344

[113] Di Paola D et al. Aflatoxin B1 toxicity in zebrafish larva (*Danio rerio*): Protective role of *Hericium erinaceus*. Toxins (Basel). 2021;**13**(10):710

[114] Guilherme S et al. Genotoxicity evaluation of the herbicide Garlon® and its active ingredient (triclopyr) in fish (*Anguilla anguilla L.*) using the comet assay. Environmental Toxicology. 2015;**30**(9):1073-1081

[115] Gupta P, Verma SK. Evaluation of genotoxicity induced by herbicide pendimethalin in fresh water fish *Clarias batrachus* (Linn.) and possible role of oxidative stress in induced DNA damage. Drug and Chemical Toxicology. 2022;**45**(2):750-759

[116] Gupta P, Verma SK. Impacts of herbicide pendimethalin on sex steroid level, plasma vitellogenin concentration and aromatase activity in teleost *Clarias batrachus* (Linnaeus). Environmental Toxicology and Pharmacology. 2020;**75**:103324

[117] de Arcaute CR, Soloneski S, Larramendy ML. Evaluation of

the genotoxicity of a herbicide formulation containing 3, 6-dichloro-2-metoxybenzoic acid (dicamba) in circulating blood cells of the tropical fish *Cnesterodon decemmaculatus*. Mutation Research/Genetic Toxicology and Environmental Mutagenesis. 2014;**773**:1-8

[118] Piancini LDS et al. Mesotrione herbicide promotes biochemical changes and DNA damage in two fish species. Toxicology Reports. 2015;**2**:1157-1163

[119] Cong B et al. The impact on antioxidant enzyme activity and related gene expression following adult zebrafish (*Danio rerio*) exposure to dimethyl phthalate. Animals (Basel). 2020;**10**(4):717

[120] Derikvandy A et al. Genotoxicity and oxidative damage in zebrafish (*Danio rerio*) after exposure to effluent from ethyl alcohol industry. Chemosphere. 2020;**251**:126609

[121] Renick VC et al. Effects of a pesticide and a parasite on neurological, endocrine, and behavioral responses of an estuarine fish. Aquatic Toxicology. 2016;**170**:335-343

[122] Shiry N et al. Exploring the combined interplays: Effects of cypermethrin and microplastic exposure on the survival and antioxidant physiology of *Astacus leptodactylus*. Journal of Contaminant Hydrology. 2023;**259**:104257

[123] Aliko A et al. Discovery of novel potential reversible peptidyl arginine deiminase inhibitor. International Journal of Molecular Sciences. 2019;**20**(9):2174

[124] Yalsuyi AM et al. Behavior evaluation of rainbow trout (*Oncorhynchus mykiss*) following temperature and ammonia alterations. Environmental Toxicology and Pharmacology. 2021;**86**:103648

[125] Faria M et al. Androgenic activation, impairment of the monoaminergic system and altered behavior in zebrafish larvae exposed to environmental concentrations of fenitrothion. Science of the Total Environment. 2021;**775**:145671

[126] Syromyatnikov MY et al. The effect of pesticides on the microbiome of animals. Agriculture. 2020;**10**(3):79

[127] Kan H et al. Correlations of gut microbial community shift with hepatic damage and growth inhibition of *Carassius auratus* induced by pentachlorophenol exposure. Environmental Science and Technology. 2015;**49**(19):11894-11902

[128] Chang X et al. Impact of chronic exposure to trichlorfon on intestinal barrier, oxidative stress, inflammatory response and intestinal microbiome in common carp *(Cyprinus carpio L.)*. Environmental Pollution. 2020;**259**: 113846

[129] Zhao H et al. Effects of environmentally relevant cypermethrin and sulfamethoxazole on intestinal health, microbiome, and liver metabolism in grass carp. Aquatic Toxicology. 2023;**265**:106760

[130] Tang J et al. Diazinon exposure produces histological damage, oxidative stress, immune disorders and gut microbiota dysbiosis in crucian carp *(Carassius auratus gibelio)*. Environmental Pollution. 2021;**269**:116129

[131] Impellitteri F et al. Evaluating quaternium-15 effects on Mytilus galloprovincialis: New insights on physiological and cellular responses.

Science of the Total Environment. 2024;**918**:170568

[132] Ramya S et al. Ecotoxicological insights: Effects of pesticides on ionic metabolism regulation in freshwater catfish, *Mystus keletius*. Aquatic Toxicology. 2023;**265**:106764

[133] Mijošek T et al. Evaluation of multi-biomarker response in fish intestine as an initial indication of anthropogenic impact in the aquatic karst environment. Science of the Total Environment. 2019;**660**:1079-1090

Chapter 11

Oxidative Stress and Exposure to Metals

Mohammad Amin Rajizadeh and Reza Pourbabaki

Abstract

Toxic metals (lead, cadmium, mercury, and arsenic) are widely found in our environment. Humans are exposed to these metals from numerous sources, including contaminated air, water, soil, and food. Recent studies have indicated that transition metals act as catalysts in the oxidative reactions of biological macromolecules. Therefore, the toxicities associated with these metals might be due to oxidative tissue damage. Redox-active metals, such as iron, copper, and chromium, undergo redox cycling, whereas redox-inactive metals, such as lead, cadmium, mercury, and others, deplete cells' major antioxidants, particularly thiol-containing antioxidants and enzymes. This chapter focuses on the relation between oxidative stress and metals exposure.

Keywords: oxidative stress, exposure, metals, mechanism, biochemical responses, physiological effects

1. Introduction

Concerning environmental and occupational health, exposure to toxic metals is a significant issue. Lead (Pb), cadmium (Cd), mercury (Hg), and arsenic (As) are widely distributed in the environment. These metals can enter human bodies through contaminated air, water, soil, or food [1, 2]. These metals are classified as redox-active or redox-inactive based on their ability to participate in redox reactions. Redox-active metals, like iron (Fe), copper (Cu) [3], and chromium (Cr), can engage in redox reactions, redox cycling, and the production of reactive oxygen species (ROS). On the other hand, redox-inactive metals like lead and cadmium can cause oxidative stress by depleting of cellular antioxidants, especially thiol-containing compounds such as glutathione (GSH) [4, 5].

2. Type of metals and their properties

2.1 Redox-active metals

Essential metals such as Fe, Cu, Cr, and cobalt (Co) have a dual nature. While they play crucial roles in many biological processes, their ability to switch between oxidation states easily (redox cycling) can have harmful consequences. These redox reactions can generate harmful molecules called ROS, especially the superoxide anion radical (O_2^-) and the highly destructive hydroxyl radical ($^{\bullet}HO$) [6]. Inside cells, these ROS act like miniature vandals, causing damage to important cellular components such as lipids, proteins, and even DNA [7–10]. The damage caused by ROS can disrupt normal cellular function and contribute to various diseases [11].

2.2 Redox-inactive metals

While redox-active metals directly participate in electron transfer reactions (redox cycling), leading to the generation of ROS, redox-inactive metals such as Cd, As, and Pb induce oxidative stress through different mechanisms [4]. These metals do not have the ability to undergo redox cycling themselves. However, they can disrupt cellular homeostasis by depleting essential antioxidant defenses, particularly thiol-containing compounds like GSH. This depletion of GSH disrupts the cell's ability to neutralize ROS, resulting in oxidative stress and ultimately cellular damage [12, 13].

3. Mechanisms of metal-induced oxidative stress

3.1 Redox-active metals

Redox-active metals such as Fe, Cu, Cr, and, Co can undergo redox cycling reactions. This process leads to the production of ROS like O_2^- and $^{\bullet}HO$. These ROS have the potential to cause significant damage to cellular components such as lipids, proteins, and DNA. For example, iron can catalyze the Fenton reaction, in which it converts hydrogen peroxide (H_2O_2) into highly reactive hydroxyl radicals. These radicals can then trigger lipid peroxidation and DNA strand breaks [4, 9].

Cu, another metal that undergoes redox reactions, can also participate in similar redox cycling reactions, leading to the production of ROS that contribute to oxidative stress. The redox cycling of Cu between its Cu and Cu states can result in the formation of superoxide radicals and H_2O_2, further exacerbating oxidative damage. Chromium, especially in its hexavalent form (Cr (VI)), can also cause oxidative stress by generating ROS and depleting cellular antioxidants [14, 15].

3.2 Redox-inactive metals

Redox-inactive metals such as Cd, As, and Pb induce oxidative stress through various mechanisms. Although these metals do not directly participate in redox cycling but can deplete cellular antioxidants, particularly thiol-containing compounds like GSH. For example, cadmium can bind to sulfhydryl groups in proteins, causing disruption in their function and leading to increased ROS production [4, 16].

Arsenic, although primarily a redox-inactive metal, can induce oxidative stress by generating H_2O_2 under physiological conditions. This ROS can then participate in Fenton-like reactions, producing highly reactive hydroxyl radicals. Pb, another redox-inactive metal, can inhibit antioxidant enzymes and deplete GSH, leading to increased oxidative stress and cellular damage [17, 18].

3.3 Cellular mechanisms

The cellular mechanisms involved in metal-induced oxidative stress encompass multiple pathways. One crucial mechanism is the disruption of mitochondrial function [19, 20]. Metals can impair mitochondrial electron transport, leading to electron leakage and the formation of superoxide radical. This dysfunction of the mitochondria can result in an overproduction of ROS, thereby aggravating oxidative stress [21, 22].

Another important mechanism is the activation of NADPH oxidase, an enzyme complex that generates superoxide radicals. Metals have the ability to activate NADPH oxidase, which increases ROS production and oxidative damage. Additionally, metals can interfere with cellular signaling pathways, such as those involving nuclear transcription factors like NF-κB and AP-1, which are sensitive to redox changes [23, 24].

4. Biochemical responses to metal exposure

The biochemical responses to metal exposure have important implications for environmental and human health. Elevated levels of oxidative stress markers, alterations in enzyme activity, and changes in serum biochemical parameters can serve as early indicators of metal toxicity in aquatic organisms and potentially in humans (**Figure 1**). It is crucial to understand these biochemical responses in order to develop effective strategies for mitigating the adverse effects of metal contamination in the environment [25, 26]. The **Table 1** shows the effects and mechanisms of various heavy metals on different species.

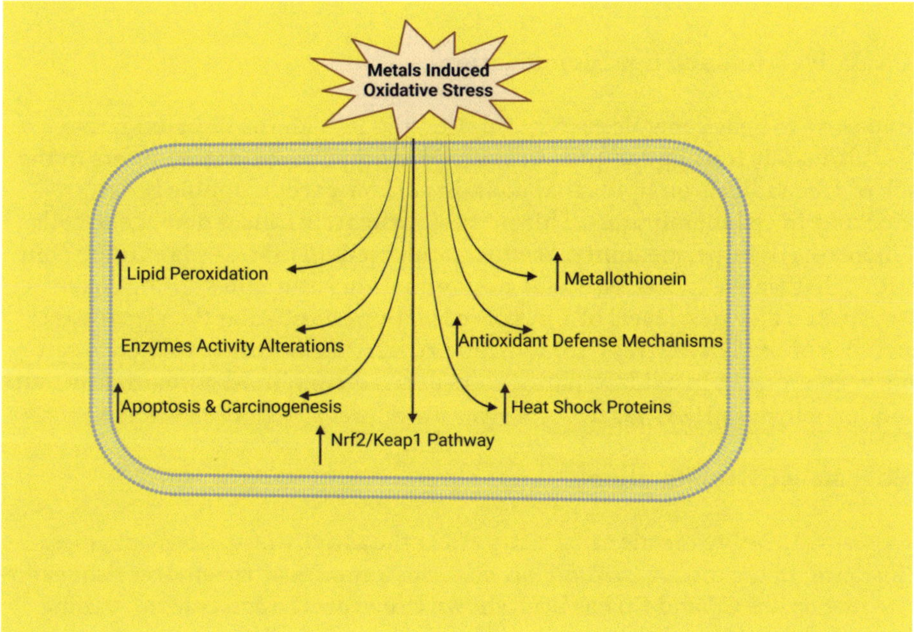

Figure 1.
The biochemical responses to metal exposure.

Metal	Disorder	Specie	Mechanism	References
Pb	Nephrotoxicity Hepatotoxicity Thyroid dysfunction Ovarian dysfunction	Mice-Rat-Human	BBB disruption-Beta amyloid deposition-Tau phosphorylation-Reduced dopamine content-Increased inflammation-Increased oxidative stress-Increased apoptosis-Increased cysts-Increased nodules	[1–7]
Hg	Autoimmune diseases Thyroid dysfunction Lung injury	Human-Sheep-Rat	Increased nodules Beta amyloid deposition Reduced dopamine content-Increased inflammation-Increased oxidative stress	[8–13]
Al	Lung injury Neurotoxicity Nephrotoxicity Hepatotoxicity	Rat	Increased inflammation-Increased oxidative stress	[14–16]
Cu	Lung injury Thyroid dysfunction	Rat	Increased inflammation-Increased oxidative stress Increased nodules	[17, 18]
Co	Neurotoxicity Lung injury Nephrotoxicity Hepatotoxicity	Rat	Increased inflammation-Increased oxidative stress Increased apoptosis	[19–22]
Cd	Ovarian dysfunction Lung injury	Rat	Increased cysts	[12, 23, 24]

Table 1.
The effects and mechanisms of various heavy metals on different species.

4.1 Oxidative stress and lipid peroxidation

Exposure to heavy metals such as Cd, Cu, and Zn can result in oxidative stress, ultimately leading to lipid peroxidation. Lipid peroxidation refers to the attack of free radicals on lipids that contain carbon-carbon double bond(s), particularly in cell membranes. This process ultimately causes damage to cells and increases their permeability. Studies conducted on marine planktonic communities that were exposed to metal contamination from mine tailings have demonstrated elevated levels of lipid peroxidation, indicating the significant occurrence of oxidative stress. It has been proven that exposure to metals can result in an increase in oxidant parameters derived from lipid peroxidation, such as Malondialdehyde (MDA) [27].

4.2 Enzyme activity alterations

Exposure to heavy metals can greatly affect the activity of various enzymes. For instance, in the case of *Oreochromis niloticus*, a species of freshwater fish, exposure to metals like Cd and Cu has been shown to elevate the activities of alanine aminotransferase [3] and aspartate aminotransferase [27]. These enzymes are recognized indicators of liver damage [28]. Additionally, fish exposed to Zn and Cd displayed higher levels of alkaline phosphatase (ALP) activity, indicating potential harm to the liver and other tissues [29].

4.3 Metallothionein induction

Metallothioneins (MTs) are low molecular weight proteins that are rich in cysteine. They have the ability to bind heavy metals and are crucial for maintaining metal detoxification and homeostasis. The production of MTs is a common biochemical response to metal exposure. For instance, when clams are exposed to treated municipal effluents containing heavy metals, there is a significant increase in MTs production. This indicates that the clams are adapting to the exposure by detoxifying and isolating the metals [30].

4.4 Changes in serum biochemical parameters

Serum biochemical parameters can serve as sensitive indicators of metal toxicity. Studies have shown that exposure to heavy metals can cause changes in serum parameters, including glucose, total protein, and cholesterol levels, in fish. For example, exposure to metals like Ag, Cd, and Cu can lead to increased glucose and total protein levels, while cholesterol levels may vary depending on the specific metal and duration of exposure. These changes are indicative of metabolic disruptions and stress responses in the organism [29].

4.5 Antioxidant defense mechanisms

The body's antioxidant defense mechanisms are crucial in reducing oxidative stress caused by exposure to metals. Enzymes such as superoxide dismutase (SOD) and catalase (CAT) play an essential role in neutralizing ROS [31]. When plants are exposed to Cu and arsenate, their SOD activity increases, indicating an upregulation of antioxidant defenses to counteract oxidative stress. Similarly, clams exposed to heavy metals show elevated CAT activity, further supporting the role of antioxidant enzymes in managing oxidative stress. SOD is not easily affected by environmental factors. Therefore, the increase in SOD activity in the serum of people living in polluted areas is one of the defense mechanisms against oxidative stress caused by environmental Pb and Cd exposure [32].

4.6 Apoptosis and carcinogenesis

Prolonged exposure to certain metals can result in programmed cell death (apoptosis) and the development of cancer (carcinogenesis). Metals like arsenic and chromium have been linked to the production of ROS, which can cause damage to DNA, promote the formation of tumors, and trigger apoptosis. The oxidative stress caused by these metals plays a crucial role in their ability to cause cancer, underscoring the importance of understanding the biochemical pathways involved in metal-induced carcinogenesis. In summary, the events that collectively lead to apoptosis during As toxicity include the generation of ROS, the accumulation of calcium ions, the upregulation of caspase-3, the downregulation of Bcl-2, and the deficiency of p53 [33].

Metals can drive tumorigenesis through various mechanisms. They can damage cellular organelles and other components, disrupt metabolic enzymes involved in detoxification, disturb cellular redox homeostasis, and cause uncontrolled oxidation. Additionally, they interfere with cell cycle regulation and cell growth, disrupt DNA repair pathways, induce DNA damage, and impact cell apoptosis and autophagy [34].

4.7 Heat shock proteins and stress response

Heat shock proteins (HSPs) are a group of proteins that are upregulated in response to stress, including exposure to metals. In sea urchin embryos exposed to heavy metals, observed rapid changes in the levels and phosphorylation states of heat shock cognate 70 (HSC70), as well as increased activity of heat shock transcription factors (HSF) and metal transcription factors (MTF). These proteins play a crucial role in protecting cells from stress-induced damage and maintaining cellular homeostasis [35].

5. Metals and signal transduction

Metals can impact the gene transcription, expression, and activation of various signaling proteins. These proteins include growth factor receptors, G-proteins like ras, tyrosine kinases such as c-src, MAPK proteins, and nuclear transcription factors like NF-κB, NFAT, AP-1, p53, and HIF-1. The effects of metals can be either activation or inactivation. These effects can occur directly through the interaction of metals with proteins or indirectly through the generation of metal-induced ROS. Moreover, the effects of metals on signaling pathways can resemble those caused by extracellular ligands like insulin or physical conditions like hypoxia [36].

5.1 Nrf2/Keap1 pathway

It is not surprising that exposure to metals activates the Nrf2 pathway. This is because several metals are known to generate ROS, and mammalian cells respond to ROS by activating Nrf2-mediated transcription. Nrf2 activation has been observed in response to various metals. Metal exposure has been found to affect the Nrf2 pathway in multiple ways. For example, it reduces sulfhydryl groups in Keap-1, activates MAPK and phosphorylates Nrf2, and inhibits proteasomal pathways, resulting in the stabilization of Nrf2. Consequently, Nrf2 is stabilized and activated, leading to the upregulation of antioxidant genes. Previous studies have demonstrated that metals induce the Nrf2-mediated oxidative stress response. However, these studies have used different exposure conditions and various human cell lines, including skin fibroblasts, hepatoma cells, monocytes, and retinal pigment epithelial cells. The use of different cell types and experimental endpoints can contribute to the variability in the oxidative stress response [37].

5.2 PI3K and Akt pathway

The metals As, Cd, Co, and Cu have been shown to interact with the PI3K/Akt pathway through ROS and non-ROS-mediated pathways [38]. Depending on the cell line involved, As may either induce or inhibit PI3K and Akt [39]. In the MCF7 breast cancer cell line, the phosphorylation of p53 induced by Cd chloride was partially mediated by PI3K [40]. Co chloride induced VEGF transcription in a dose- and time-dependent manner *via* the PI3K pathway [41].

5.3 MAPK pathway

Several metals have been found to impact MAPK signaling. These metals include As, Cr (VI), Cd, Co, and Ni [36]. As (III) has been shown to strongly activate ERK1 and ERK2, JNK, and p38 in a human bronchial epithelial cell line [42]. Cd stimulates the activation of HO-1 through JNK and p38 in CL3 cells and through p38 in the MCF-7 human breast cancer cell line [43].

5.4 NFκB pathway

Recent studies have shown that various metal ions can impact the activation and activity of NF-κB. These metal ions include As (III), Cr (VI), Ni, Co, Cd, Pb, and Fe. As (III) has been found to influence NF-κB, acting as both an inducer and an inhibitor in different scenarios [36].

6. Physiological effects of metal-induced oxidative stress

The **Figure 2** shows that oxidative stress can lead to a variety of health problems. The severity of these health problems can vary depending on the type of metal, the level of exposure, and the individual's susceptibility.

6.1 Impact on the nervous system

One of the most significant impacts of metal-induced oxidative stress is observed in the nervous system. Metals such as aluminum (Al), Pb, and Hg have been associated with neurodegenerative diseases like Alzheimer's and Parkinson's disease. These metals can induce oxidative stress by disrupting mitochondrial function and increasing the production of ROS, leading to neuronal damage and apoptosis. For instance, exposure to Al has been shown to impair mitochondrial bioenergetics, which contributes to the development of neurodegenerative disorders [22].

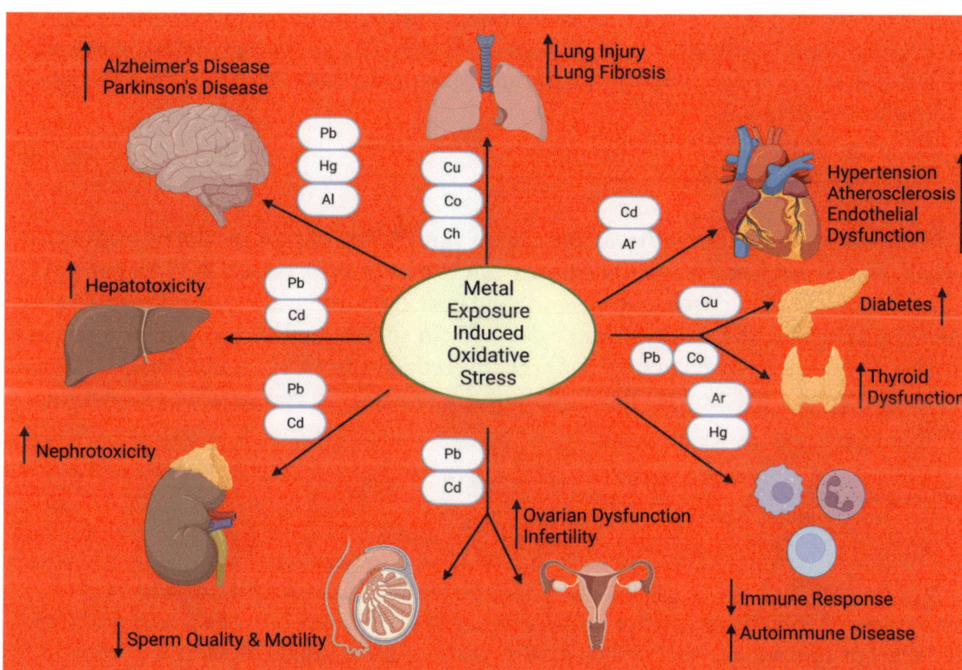

Figure 2.
Oxidative stress can lead to a variety of health problems.

6.2 Cardiovascular effects

Heavy metals can have harmful effects on the cardiovascular system. For instance, metals such as arsenic and Cd are known to cause oxidative stress. This oxidative stress can ultimately lead to endothelial dysfunction, atherosclerosis, and hypertension. These metals can harm the endothelial cells that line the blood vessels and promote inflammation and plaque formation. Additionally, oxidative stress can have a negative impact on nitric oxide, a crucial molecule for vascular health, thereby exacerbating cardiovascular problems [44].

6.3 Hepatotoxicity

The liver plays a crucial role in detoxification, making it particularly susceptible to metal-induced oxidative stress. Metals like Cd and Pb tend to accumulate in the liver, which can result in heightened production of ROS and subsequent oxidative damage. This can lead to lipid peroxidation, protein oxidation, and DNA damage, ultimately resulting in liver dysfunction and hepatotoxicity. Numerous studies have shown that exposure to these metals can drastically affect liver enzyme activities, indicating liver damage and impaired metabolic functions [45].

6.4 Renal effects

The kidneys are highly sensitive organs that can be negatively impacted by oxidative stress caused by metal exposure. Metals like Cd and Pb have the ability to build up in the kidneys, resulting in oxidative damage and nephrotoxicity. This oxidative stress can impair renal function by harming the glomeruli and tubules, ultimately leading to proteinuria, a decrease in glomerular filtration rate, and the development of chronic kidney disease. Furthermore, the oxidative stress in the kidneys can also initiate inflammatory responses, exacerbating the damage to the renal system [46].

6.5 Reproductive toxicity

Metal-induced oxidative stress can have a significant impact on the reproductive system. Metals such as Pb and Cd can disrupt the balance between ROS and antioxidants in reproductive tissues, leading to oxidative damage and impaired reproductive function. In males, this can result in a decrease in sperm quality and motility, while in females, it can cause ovarian dysfunction and reduced fertility. Additionally, oxidative stress can also affect fetal development, increasing the risk of abnormalities and miscarriage [47].

6.6 Immunotoxicity

The immune system is vulnerable to oxidative stress induced by metals. Metals such as As and Hg have the ability to disrupt immune function by causing oxidative stress and interfering with the balance between pro-oxidants and antioxidants. This disruption can weaken immune cells, making individuals more susceptible to infections and reducing their immune response. Additionally, oxidative stress can also trigger chronic inflammation, which in turn plays a role in the development of autoimmune diseases and other immune-related disorders [48].

6.7 Pulmonary toxicity

Metal particles that are inhaled or ingested can lead to pulmonary fibrosis and functional impairment. The severity of these conditions depends on the fibrogenic potential of the metal agent [36]. In the context of ongoing pulmonary injury and inflammation leading to fibrosis, metal-related interstitial lung diseases can be categorized into several groups:

1. Diffuse fibrotic interstitial lung diseases, which are commonly observed in patients exposed to uranium and arsenic.

2. Benign pneumoconiosis, which involves minimal or no fibrosis, such as siderosis caused by inhalation of iron compounds.

3. Sarcoid-like or hypersensitivity pneumonitis-like epithelioid granulomatous lung diseases, reported with exposure to beryllium, aluminum, and titanium.

4. Giant cell and desquamative interstitial pneumonitis, observed in workers exposed to Co and tungsten.

5. Acute chemical pneumonitis, seen in cases of aluminum, cadmium, and nickel exposure [49].

6.8 Diabetes

Ceruloplasmin levels have been found to be higher in individuals with both type 1 and type 2 diabetes compared to healthy individuals. Furthermore, several studies have reported increased concentrations of Cu in the plasma of diabetic patients who have complications such as hypertension and retinopathy. This altered Cu metabolism, combined with elevated levels of glycated proteins, may contribute to the development of complications related to diabetes. Glycated proteins have a stronger attraction to transition metal ions, including Cu. Even though Cu is bound to proteins, it can still contribute to the formation of free radicals, leading to increased oxidative stress in individuals with diabetes. Various markers of oxidative damage, such as damaged proteins, lipid peroxidation, and DNA damage, have been observed and implicated in the development of diabetic complications [50].

7. Therapeutic approaches

Therapeutic approaches to mitigate metal-induced oxidative stress include the use of chelating agents and antioxidant supplementation. Chelation therapy involves the administration of chelating agents such as calcium disodium ethylenediamine tetraacetic acid (CaNa2EDTA), British Anti-Lewisite (BAL), and meso-2,3-dimercaptosuccinic acid (DMSA), which bind to metal ions and facilitate their excretion from the body. However, chelation therapy can have serious side effects, and its efficacy is often limited [51].

Antioxidants may be used as a supportive therapy to chelation therapy. Studies show that antioxidants can help mitigate the oxidative damage caused by metal exposure. For example, vitamin E and melatonin have been demonstrated to protect against oxidative damage induced by metals *in vitro* and *in vivo*. Combining antioxidants with chelating agents may provide a potentially more efficacious treatment approach by reducing metal burden and oxidative stress simultaneously [52].

8. Conclusion

Metal exposure, particularly to redox-active and redox-inactive metals, disrupts cellular homeostasis and leads to oxidative stress. Redox-active metals such as Fe, Cu, and Cr generate ROS through redox cycling, while redox-inactive metals like Cd, Pb, and As deplete cellular antioxidants, creating an environment conducive to ROS production. This oxidative stress can cause significant damage to cellular components, including lipids, proteins, and DNA, ultimately contributing to various diseases. Understanding these mechanisms is crucial for developing therapeutic strategies to combat metal-induced oxidative stress. Chelation therapy and antioxidant supplementation are potential approaches, but they have limitations and require further research to optimize their efficacy.

Acknowledgements

I acknowledge Publishing Process Manager Tea Jelaca at IntechOpen for guiding me through the editorial process. AI-powered language tools were used to assist with grammar and style checking, but all content was generated by the authors.

Conflict of interest

The authors declare no conflict of interest.

Author details

Mohammad Amin Rajizadeh[1] and Reza Pourbabaki[2*]

1 Physiology Research Center, Institute of Neuropharmacology, Kerman University of Medical Sciences, Kerman, Iran

2 Department of Occupational Health, School of Health, Shiraz University of Medical Sciences, Shiraz, Iran

*Address all correspondence to: r.pourbabaki@gmail.com

References

[1] Quist AJ et al. Metal exposures in residents living near an urban oil drilling site in Los Angeles, California. Environmental Science and Technology. 2022;**56**(22):15981-15989

[2] Wrzecińska M et al. Disorders of the reproductive health of cattle as a response to exposure to toxic metals. Biology. 2021;**10**(9):882

[3] Safety O, H. Administration. Pocket Guide to Chemical Hazards. Cincinnati: OSHA; 2014

[4] Valko M et al. Redox-and non-redox-metal-induced formation of free radicals and their role in human disease. Archives of Toxicology. 2016;**90**:1-37

[5] Chai Y-C, Mieyal JJ. Glutathione and glutaredoxin—Key players in cellular redox homeostasis and signaling. Antioxidants. 2023;**12**(8):1553

[6] Bejeshk MA et al. Perillyl alcohol (PA) mitigates inflammatory, oxidative, and histopathological consequences of allergic asthma in rats. Naunyn-Schmiedeberg's Archives of Pharmacology. 2023;**396**(6):1235-1245

[7] Ikeda Y, Nagasaki Y. Antioxidative biointerface: Biocompatible materials scavenging reactive oxygen species. Biomedical Materials. 2018;**13**(4):044103

[8] Larosa V, Remacle C. Insights into the respiratory chain and oxidative stress. Bioscience Reports. 2018;**38**(5):BSR20171492

[9] Li W et al. Generation of reactive oxygen species by promoting the Cu (II)/Cu (I) redox cycle with reducing agents in aerobic aqueous solution. Water Science and Technology. 2018;**78**(6):1390-1399

[10] Shakerinasab N et al. The hydroalcoholic extract of nasturtium officinale reduces lung inflammation and oxidative stress in an ovalbumin-induced rat model of asthma. Evidence-based Complementary and Alternative Medicine. 2022;**2022**(1):5319237

[11] Zuo L, Wijegunawardana D. Redox role of ROS and inflammation in pulmonary diseases. Lung Inflammation in Health and Disease. 2021;**II**:187-204

[12] Aoyama K, Nakaki T. Glutathione in cellular redox homeostasis: Association with the excitatory amino acid carrier 1 (EAAC1). Molecules. 2015;**20**(5):8742-8758

[13] Habermann KJ et al. Targeting redox homeostasis in rhabdomyosarcoma cells: GSH-depleting agents enhance auranofin-induced cell death. Cell Death and Disease. 2017;**8**(10):e3067-e3067

[14] Rehman SU et al. Redox cycling of Cu (II) by 6-mercaptopurine leads to ROS generation and DNA breakage: Possible mechanism of anticancer activity. Tumor Biology. 2015;**36**:1237-1244

[15] Cheignon Cm et al. Oxidative stress and the amyloid beta peptide in Alzheimer's disease. Redox Biology. 2018;**14**:450-464

[16] Nowicka B et al. The inhibitor-evoked shortage of tocopherol and plastoquinol is compensated by other antioxidant mechanisms in *Chlamydomonas reinhardtii* exposed to toxic concentrations of cadmium and chromium ions. Ecotoxicology and Environmental Safety. 2020;**191**:110241

[17] Lee G et al. Copper arsenite-complexed fenton-like nanoparticles as oxidative stress-amplifying anticancer agents. Journal of Controlled Release. 2022;**341**:646-660

[18] Zhang Z et al. The protective effect of baicalin against lead-induced renal oxidative damage in mice. Biological Trace Element Research. 2017;**175**:129-135

[19] Emami MH et al. Nrf2 signaling pathway in trace metal carcinogenesis: A cross-talk between oxidative stress and angiogenesis. Comparative Biochemistry and Physiology Part C: Toxicology and Pharmacology. 2022;**254**:109266

[20] Kumar S et al. Genetic alterations and oxidative stress in T cell lymphomas. Pharmacology and Therapeutics. 2022;**236**:108109

[21] Belyaeva EA et al. Mitochondrial electron transport chain in heavy metal-induced neurotoxicity: Effects of cadmium, mercury, and copper. The Scientific World Journal. 2012;**2012**(1):136063

[22] Cheng H et al. Mechanisms of metal-induced mitochondrial dysfunction in neurological disorders. Toxics. 2021;**9**(6):142

[23] Ray SK et al. Oxidative versus reductive stress in breast cancer development and cellular mechanism of alleviation: A current perspective with anti-breast cancer drug resistance. Current Molecular Medicine. 2024;**24**(2):205-216

[24] Filip-Ciubotaru F et al. NADPH oxidase: Structure and activation mechanisms (review). Note I. The Medical-Surgical Journal. 2016;**120**(1):29-33

[25] Ma C et al. Metal-based nanotoxicity and detoxification pathways in higher plants. Environmental Science and Technology. 2015;**49**(12):7109-7122

[26] Pourbabaki R et al. A review on antioxidant-rich products as protective and therapeutic agents against chlorpyrifos hepatotoxicity. Iran Occupational Health. 2021;**18**(1):114-138

[27] Fernández-Ramos MD et al. An optical sensor for the sensitive determination of formaldehyde gas based on chromotropic acid and 4-aminoazobenzene immobilized in a hydrophilic membrane. Analyst. 2023;**148**(18):4533-4538

[28] Zhou X, Feng Y, Gong Z. Associations between lead, cadmium, mercury, and arsenic exposure and alanine aminotransferase elevation in the general adult population: An exposure–response analysis. Environmental Science and Pollution Research. 2022;**29**(35):53633-53641

[29] Shahjahan M et al. Effects of heavy metals on fish physiology–A review. Chemosphere. 2022;**300**:134519

[30] Reddy MS et al. Metal induction of a *Pisolithus albus* metallothionein and its potential involvement in heavy metal tolerance during mycorrhizal symbiosis. Environmental Microbiology. 2016;**18**(8):2446-2454

[31] Drozdz-Afelt JM, Koim-Puchowska BB, Kaminski P. Analysis of oxidative stress indicators in polish patients with prostate cancer. Environmental Science and Pollution Research. 2022;**29**:4632-4640

[32] Wieloch M et al. Do toxic heavy metals affect antioxidant defense mechanisms in humans? Ecotoxicology

and Environmental Safety. 2012;**78**:195-205

[33] Rana SVS. Metals and apoptosis: Recent developments. Journal of Trace Elements in Medicine and Biology. 2008;**22**(4):262-284

[34] Zhu Y, Costa M. Metals and molecular carcinogenesis. Carcinogenesis. 2020;**41**(9):1161-1172

[35] Mahmood K et al. Synergistic effects of toxic elements on heat shock proteins. BioMed Research International. 2014;**2014**(1):564136

[36] Leonard SS, Harris GK, Shi X. Metal-induced oxidative stress and signal transduction. Free Radical Biology and Medicine. 2004;**37**(12):1921-1942

[37] Simmons SO et al. NRF2 oxidative stress induced by heavy metals is cell type dependent. Current Chemical Genomics. 2011;**5**:1

[38] Souza K et al. Arsenite activation of PI3K/AKT cell survival pathway is mediated by p38 in cultured human keratinocytes. Molecular Medicine. 2001;**7**:767-772

[39] Choi Y-J et al. Arsenic trioxide-induced apoptosis in U937 cells involve generation of reactive oxygen species and inhibition of Akt. International Journal of Oncology. 2002;**21**(3):603-610

[40] Matsuoka M, Igisu H. Cadmium induces phosphorylation of p53 at serine 15 in MCF-7 cells. Biochemical and Biophysical Research Communications. 2001;**282**(5):1120-1125

[41] Huang G-W, Yang L-Y, Lu W-Q. Effects of PI3K and p42/p44 MAPK on overexpression of vascular endothelial growth factor in hepatocellular carcinoma. World Journal of Gastroenterology. 2004;**10**(6):809

[42] Samet JM et al. Activation of MAPKs in human bronchial epithelial cells exposed to metals. American Journal of Physiology-Lung Cellular and Molecular Physiology. 1998;**275**(3):L551-L558

[43] Alam J et al. Mechanism of heme oxygenase-1 gene activation by cadmium in MCF-7 mammary epithelial cells: Role of p38 kinase and Nrf2 transcription factor. Journal of Biological Chemistry. 2000;**275**(36):27694-27702

[44] Balakumar P, Kaur J. Arsenic exposure and cardiovascular disorders: An overview. Cardiovascular Toxicology. 2009;**9**:169-176

[45] Andjelkovic M et al. Toxic effect of acute cadmium and lead exposure in rat blood, liver, and kidney. International Journal of Environmental Research and Public Health. 2019;**16**(2):274

[46] Elinder C-G, Barregard L. Renal effects of exposure to metals. In: Handbook on the Toxicology of Metals. Netherlands: Elsevier; 2022. pp. 485-506

[47] Patwa J, Sharma A, Flora S. Arsenic, cadmium, and lead. In: Gupta RC, editor. Reproductive and Developmental Toxicology. 3nd ed. Netherlands: Elsevier; 2022. pp. 547-571

[48] Calderón-Garcidueñas L et al. Immunotoxicity and environment: Immunodysregulation and systemic inflammation in children. Toxicologic Pathology. 2009;**37**(2):161-169

[49] Assad N et al. Metal-induced pulmonary fibrosis. Current Environmental Health Reports. 2018;**5**:486-498

[50] Jomova K, Valko M. Advances in metal-induced oxidative stress and human disease. Toxicology. 2011;**283**(2-3):65-87

[51] Flora S, Mittal M, Mehta A. Heavy metal induced oxidative stress and its possible reversal by chelation therapy. Indian Journal of Medical Research. 2008;**128**(4):501-523

[52] Amadi CN et al. Natural antidotes and management of metal toxicity. Environmental Science and Pollution Research. 2019;**26**:18032-18052